Ensuring the Durability of Oil-Producing Pumps Through the Use of Laser Spraying Technology

Vitaliy Vladimirovich SAVINKIN

Olga Vladimirovna IVANOVA

Andrei Victor SANDU

Sergey Nikolaevich KOLISNICHENKO

Reviewers:
Doctor of Technical Sciences Kuznetsova V.N., Dean of the Faculty of Oil and Gas and Construction Equipment (Siberian State Automobile and Road Engineering University, Omsk, Russia);
Candidate of Technical Sciences S.B. Musrepov, Director (North Kazakhstan College);
PhD A.A. Kashevkin, Head of Chair, Associate Professor of department of energetic and radioelectronics (Non-profit limited company «Manash Kozybayev North Kazakhstan university»)

Published by **Materials Research Forum LLC**
Millersville, PA 17551, USA

Published as part of the book series
Materials Research Foundations
Volume 144 (2023)
ISSN 2471-8890 (Print)
ISSN 2471-8904 (Online)

Print ISBN 978-1-64490-234-9
ePDF ISBN 978-1-64490-235-6

Distributed worldwide by

Materials Research Forum LLC
105 Springdale Lane
Millersville, PA 17551
USA
http://www.mrforum.com

Printed in the United States of America
10 9 8 7 6 5 4 3 2 1

Table of Contents

Preface... *v*

Designations and Abbreviations.. *v*

Introduction 1

**Chapter 1. Studies of the design features in pumping complexes during
operation in an aggressive environment of an oil fluid**............................5

   1.1   Oil-producing technologies and applied
oil-producing equipment 5

   1.2   Electric centrifugal pumping plants 6

   1.3   Advantages and disadvantages of electric centrifugal pumps 11

   1.4   Rod depth pumping plant..14

   1.5   Design features of the rod depth pump...15

   1.6   Main details of the rod depth pump 20

   1.7   Patents of non-standard pump designs...24

   1.8   Causes of malfunctions of rod depth pumps......................................26

   1.9   Methodology for calculating the dosage of salt
deposition inhibitors...31

   1.10   Methods of restoring pump operability...37

   Conclusion..39

Chapter 2. Gas dynamic analysis of a downhole rod deep pump............41

   2.1   Overview of valve pair throughput 41

   2.2   Preliminary calculation of RDP operating parameters 43

   2.3   Calculation of RDP performance indicators in the production
of multicomponent oil...51

   2.4   Research of modern technologies of recovery of
mining pumps..60

   Conclusion..66

**Chapter 3. Development of production-technological equipment for
special purposes** 69

   3.1   Analysis of existing technological equipment 69

   3.2   Calculation of cutting modes satisfying the effective
operation of the developed boring head..71

3.3   Strength calculation of the body of the developed
boring head 83

Conclusion 88

**Chapter 4. Controlled installation for laser spraying of internal surfaces
of small diameter** **90**

4.1   Description of a controlled installation for laser spraying of
internal surfaces of small diameter 90

4.2   Components of the developed spray head 92

4.3   Development of a management system 96

4.4   Calculation of optimal characteristics of the laser source 100

4.5   Development of technology for manufacturing a rod pump
cylinder using a laser spraying unit 105

Conclusion 109

Conclusion **110**

References **111**

Preface

The monograph examines and investigates the issues of trouble-free operation of producing pumping units, the main causes of failures and wear of mechanisms as a result of the aggressive action of the oil-producing environment. The results of the gas-dynamic calculation of the main loaded elements of the pump using Simulation Flow software based on the CAD editor of SolidWorks for the analysis and visualization of processes occurring in the working space of the rod deep pump are presented. To ensure the durability of oil-producing pumps, the application of laser spraying technology is proposed with a description of a unique controlled installation for laser modification of the internal surfaces of critical and highly loaded parts of the studied small-diameter structures.

The monograph is intended for bachelors, masters, doctoral students and specialists in the field of mechanical engineering and energy, and can also be used by employees of scientific, design, production and repair and restoration enterprises.

Designations and Abbreviations

PDRP – plant of downhole rod pump
ECPP – electric centrifugal pump plant
ECPPC – electric centrifugal pump plant with increased corrosion wear resistance
ECP – electric centrifugal pump
RDP – rod depth pump
FCP – filtration-capacitance properties
PCP – pumping-compressor pipes
NP – the non-plug-in pump
PP – the plug-in borehole pump
ARPD – asphalt-resin-paraffin deposits
RRW – the routine repair of wells
OHP – overhaul period

Introduction

The oil and gas industry plays an important role in the development of the economy of any dynamically developing country. To date, of the huge number of oil and gas wells around the world, only about 5% are gushing. According to experts, the fund of small and medium-sized wells (with a flow rate of less than 30 m^3/day) currently accounts for more than 50% of the total number of wells in operation. First of all, this is due to a decrease in the reserves of "light" oil due to its intensive production and depletion of existing fields. At the same time, the number of newly discovered deposits of light and medium oils is constantly decreasing. Currently, there is a tendency to increase the share of reserves of hard-to-recover high-viscosity oil [1, 2]. Oil-producing companies face a serious problem related to the increase in hydrocarbon production not only due to the exploration and development of new fields, but also due to the renewal of the technical park of oil-producing equipment and the use of modern technologies and methods of oil production. Deposits with low- and medium-rate wells are in most cases equipped with plant of downhole rod pumps (PDRP), low-rate (supply up to 30 m^3/day) electric centrifugal pumps (ECP), electric centrifugal pump plants (ECPP) with periodic operation, screw pumps and systems of simultaneous separate operation [3 – 11].

As a rule, the process of oil extraction is realized with the use of specialized equipment and lifting-power units operating under extreme operating conditions. The scientific-technical problem lies in the fact that the pumping unit is put into operation at the specified factory (ideal) parameters. Since the beginning of operation of oil-producing pumps, their technical characteristics have been changing and progressively do not correspond to the nominal values set in laboratory conditions at the manufacturer. The wear of mechanisms, as a result of the aggressive action of the environment, reduces the service life of mining plants. As a result, the need to increase the resource of oil production facilities increases dramatically, which requires new technologies for manufacturing critical and highly loaded parts of the structures under study.

The significance of the study of the problem of increasing the productivity and operability of structures of oil-producing equipment is directly proportional to the importance of oil production for humanity as a whole.

Therefore, the study of existing oil pump designs with the analysis of design and technological features and the technological solutions used will solve practical problems of increasing productivity, working life and production capacity of the well, increasing the inter-repair period of the oil pump as a result of the application of innovative methods of manufacturing oil-producing equipment.

One of the methods of studying the design and technological features of pumping equipment is the method of modeling hydrodynamic processes occurring in the pump in real time.

The main purpose of modeling hydrodynamic processes in an oil-producing pump is to solve the main tasks and establish the dependencies of the technological parameters of the outflow of a complex medium on the structural elements of the pump in the underground hydrodynamics of wells. These studies make it possible to determine the filtration-capacitance properties (FCP) of the reservoir directly during the operation of the oil production pump and the supply of oil fluid [12, 13]. Modern hydrodynamic research methods make it possible to obtain the most important parameters based on field data, on the basis of which field development systems are designed, the process of oil production is regulated and the efficiency of object development is analyzed.

Using the modeling of hydrodynamic processes in the Simulation Flow software environment based on the CAD editor of SolidWorks, the following tasks can be solved: 1) measurement of the well flow rate and determination of the physico-chemical properties of the extracted liquid; 2) determination of downhole and reservoir pressures in time; 3) measurement of the temperature of the extracted liquid; 4) registration of the flow velocity and fluid density at various sections of the flow; 5) and other parameters of the extracted liquid related to productive oil and gas reservoirs [12 – 15].

One of the phenomena occurring in oil and gas pumps is the phenomenon of cavitation, that is, a violation of the continuity of the oil fluid due to the appearance of air bubbles (gas, steam) caused by a pressure drop. Under the influence of high pressure, the cavities filled with gas collapse. This leads to the occurrence of a local hydraulic shock (pressure of several thousand atmospheres) and, as a result, to the destruction of the surface layer of the parts.

Modeling of hydrodynamic processes in the Simulation Flow software environment based on the CAD editor of SolidWorks allows you to determine the zones of possible occurrence of cavitation phenomena and thereby solve the problem of destruction of the surface layer of oil and gas pump parts.

Oil liquids lying on a large area of the earth are characterized by the complexity of production due to increased wear of equipment from contained mechanical impurities and highly paraffin deposits with a high gas content.

An actual solution to reduce the wear of oil-producing equipment is the use of modern methods of hardening the working surfaces of the parts of the extraction pump. Such methods include the technology of laser spraying with the use of concentrated light sources. A feature of the laser spraying technology is the possibility of creating a radiation flux in the spraying zone of sufficiently high density, which allows to obtain high temperatures (more than 3000 °C); the possibility of creating the smallest thermal impact zone (in time and area); contactless input of energy into the processing zone and others. The above features of laser radiation can be used for surface hardening of worn parts of oil-producing equipment.

In the process of investigating the problems of the rod deep pump, the main causes of failures and their further consequences on the operation of the entire oil production complex were identified. Based on the above research, it was found that more than 50% of all rod pump failures are caused by mechanical impurities associated with the fluid flow and falling into the precision gaps of the plunger-cylinder friction pairs lead to their wear.

The cylinder of the rod depth pump is one of the main elements of the pump design. Examining the details of the "plunger-cylinder", the high complexity of processing long-length solid cylinders, which are a pipe with a precision-machined inner surface, was determined. The cylinder, due to the specifics of production, dimensions, as well as the requirements imposed on it, creates the need for the use of specialized equipment and tooling. Therefore, the design and manufacture of high-quality and high-tech equipment for the pretreatment of RDP cylinders is also one of the priorities of the machine-building industry.

As is known, surface hardening, as a result of spraying metals on the surface of the product, is a technology used to change the structure and properties of the surface of the product, to ensure certain qualities that increase the maintenance period of operation of structural parts operating in an aggressive environment and exposed to high temperatures, vibration and dynamic alternating loads. Metal spraying in modern metallurgy and mechanical engineering is considered one of the most economical methods of surface treatment of products. This technology makes it possible to obtain the desired operational properties of the surface at the lowest cost.

Spraying is a very complex process that requires fine-tuning of technological equipment to obtain a high quality coating. One of the main characteristics that ensures the quality of spraying is the choice of a laser source.

The invention proposed by the authors of the project relates to the field of mechanical engineering, metallurgy and other industries, and specifically to the technique and technology of laser welding production, having a laser head, a mirror reflector, a pulse-discharge lamp, a movable carriage, a cooling system, a welding unit providing restoration and modification of a worn surface, laser welding, hardening, application of

Materials Research Forum LLC
https://doi.org/10.21741/9781644902356

protective coatings on the internal working surfaces of cylinders of oil-producing deep-rod pumps and long-length steel pipes of small diameter.

The technical result achieved by this invention consists in increasing the constructive and technological efficiency of the operation of a portable laser installation, including through automated control, adaptability, increased productivity, reduced labor intensity of work and expansion of its functionality, as well as in improving the quality of adhesion of the coating to the substrate when restoring the internal surfaces of small diameter long pipes of the pump type RDP.

Chapter 1. Studies of the design features in pumping complexes during operation in an aggressive environment of an oil fluid

1.1   Oil-producing technologies and applied oil-producing equipment

The discovery of the useful properties of such a fossil as oil marked the beginning of the rapid development of extraction methods. Methods of oil extraction are divided into fountain, gas lift and pumping equipment. Oil extraction using pumps is used when the reservoir pressure has become less than the bottom hole and oil extraction requires additional resources. The heterogeneity of the chemical composition and the content of mechanical impurities in the oil-containing liquid impose restrictions on the operational characteristics of the equipment, in connection with which pumping units of the following types have been developed:

1.   rod depth pumps (RDP);

2.   electric centrifugal pump;

3.   screw pumps;

4.   diaphragm pumps;

5.   hydraulic piston pumps.

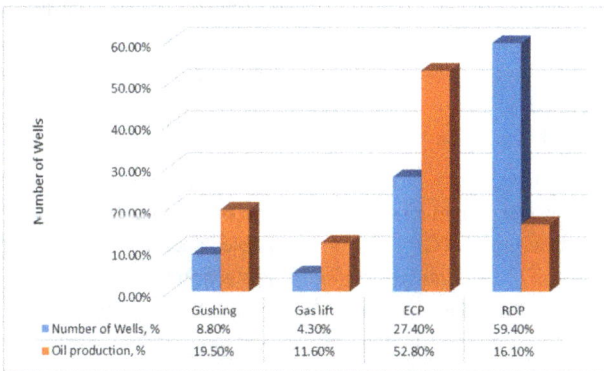

Figure 1. Crude oil extraction statistics by common extraction methods

Materials Research Foundations **144** (2023)   https://doi.org/10.21741/9781644902356

The results of statistical studies (Figure 1) have shown that most wells in the CIS countries are equipped with ECP and RDP units. RDP surpasses the ECP technology in terms of the number of wells involved, but they are more than 3 times inferior in the volume of mineral extraction. The big difference between the technologies used raises the question of the need to conduct a study of the designs of oil-producing plants.

1.2   Electric centrifugal pumping plants

The electric centrifugal pump is designed for the extraction of borehole oil-containing liquid. ECPs can be used in deep and inclined oil wells, also in conditions of heavily watered wells with high mineralization of reservoir waters and for lifting salt and acid solutions. The extracted liquid consists of oil, associated water (no more than 99%) and gas with a density of no more than 1,400 kg/m^3; temperature from 50 °C to 90 °C; solid content not more than 0.1 g/l; hydrogen sulfide concentration not more than 0.01 g/l; water acidity (pH) from 6 to 8.5; with pump outlet pressure up to 23 MPa [16, 17].

Depending on the operating conditions, namely, on the number of different components contained in the pumped liquid, plants of conventional design (ECPP) and with increased corrosion wear resistance (ECPPC) have been developed [16, 18].

With an increase in the gas content, cavities are formed in the channels of the impellers and the guiding devices of the pump that are not involved in the general flow of the mixture through the channels. The appearance of gas-filled caverns leads to a decrease in the capacity of the pump channels and a sharp deterioration in the flow of the blades, disrupting the energy exchange between the pump and the pumped medium. The pump operates in artificial cavitation modes and with a further increase in the gas content, a pump supply failure may occur. With a sharp filling of the gas cavity of the entire working area of the cylinder and the subsequent failure of the supply can lead to an overload of the electric motor. The pressure-flow curves for the mixture are located below the characteristics of the pump on a non-carbonated liquid [16 – 18].

The centrifugal pumping unit consists of a submersible working part and a ground control. The submersible part includes: an electric pump that descends into the well on pumping-compressor pipes (PCP). The submersible electric pump consists of a high-power asynchronous electric motor, hydraulic protection of the electric motor, an inlet gas separator, an assembly of turbines or working blades pumping the fluid flow, return and drain valves. The ground equipment of the ECP unit includes: electrical equipment of the installation and wellhead equipment of the well. The electrical equipment consists of a control station with a transformer and a transformer substation. The connection of the ground cable to the main line is carried out in a terminal box at a distance of 5 – 10 meters from the mouth. After the terminal box, the cable is connected along the chain to the transformer, then the control station and then through the transformer substation, the entire pumping unit receives electricity from the power plant (Figure 2) [16 – 18].

Figure 2 shows that the transformer substation converts the field network voltage to the required optimal value at the terminals of the electric motor, taking into account the voltage losses in the cable.

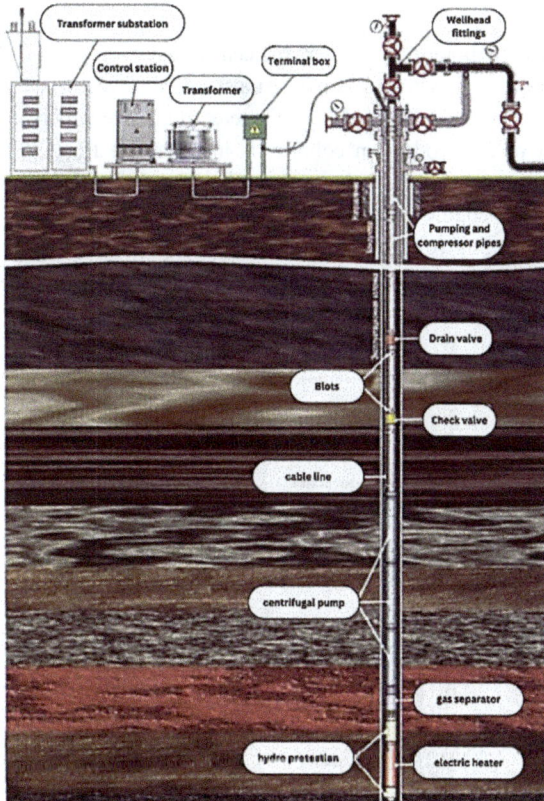

Figure 2. Scheme of an electric center pumping unit [19]

The control station provides control over the operation of pumping units and its protection under optimal conditions. Wellhead equipment is designed for sealing the mouth and regulating oil extraction. In the equipment of the wellhead, the column of pumping-compressor pipes is located eccentrically relative to the axis of the well, which allows conducting research work through the inter-tube space [19, 20].

Electricity from the transformer to the submersible electric motor is supplied via a ground cable. The connection of the ground cable to the main line is carried out in a terminal box at a distance of 5 – 10 meters from the mouth. After the terminal box, the cable is connected along the chain to the transformer – control station – transformer substation – power plant [19, 20].

Asynchronous electric motor of a submersible centrifugal pump (Figure 3) it is made in a vertical design and a steel cylindrical body. At a standard current frequency of 50 Hz, the engine shaft speed reaches 3,000 rpm. The diameter of electric motors, determined by the internal diameter of the production column, is in the range from 96 to 130 mm. The parameters of power, current and voltage depend on the size of the motor. Currently, the nominal power of the engine varies from 8 to 500 kW with an operating current from 18 to 180 A and an operating voltage from 300 to 3,600 V [19, 20].

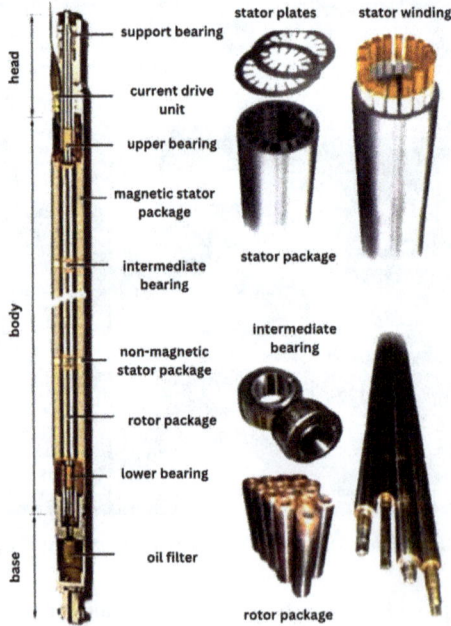

Figure 3. ECP electric motor scheme [19]

The submersible electric motor consists of a stator, a rotor, a head and a base. The stator is a stationary part of the engine. The stator housing is made in the form of a steel pipe with a thread at the ends for connecting the head and base of the electric motor. The stator consists of alternating magnetic and non-magnetic packages that are pressed into the housing. Plates of magnetic packages are stamped from electrical steel, and non-magnetic packages from brass or non-magnetic steel. Non-magnetic packages serve as supports for intermediate rotor bearings [19, 20].

The oil circulating inside the engine transfers heat to the stator and through the iron and the stator housing – the reservoir fluid washing the engine. Therefore, in order to cool the engine, a continuous flow of reservoir fluid through the annular gap between the motor housing and the production column is necessary. And the higher the rate of passage of the reservoir fluid, the better the cooling will be carried out [19, 20].

The hydraulic protection separates the reservoir fluid from the cavity where the electric motor is located (Figure 4). Hydraulic protection is installed on both sides of the electric motor to compensate for the thermal expansion of the oil and transfer the torque to the shaft of the centrifugal pump.

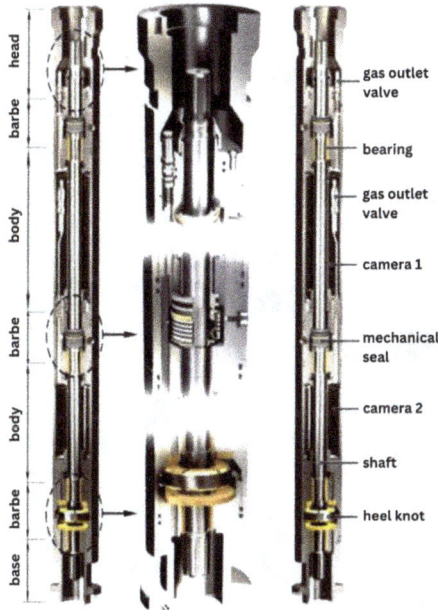

Figure 4. Hydraulic protection scheme with gas separator [21]

Gas separators are used to neutralize the negative effects of free gas in a liquid. Gas separators are most effective when the volume content of gas at the pump inlet is more than 30%. With a lower content of free gas, the negative effect does not manifest itself. During the operation of the gas separator, the flow of degassed liquid is fed to the next stage of the pump, and the gas is discharged through the separation holes into the annulus [20, 21].

The submersible multistage centrifugal pump is a set of stages of impellers and guiding devices (Figure 5). Impellers and guide devices are assembled on a single shaft, which is supported by an axial support. Centrifugal pumps are divided into three assembly schemes:

1. "floating" assembly;
2. "compression" assembly;
3. "batch" assembly.

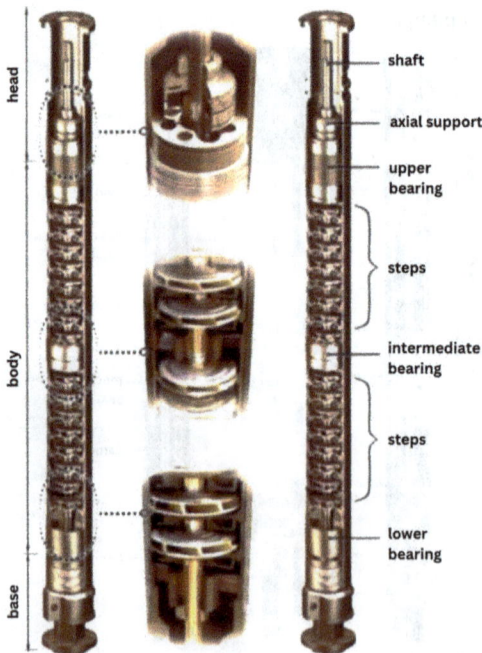

Figure 5. Assembly scheme of the stages of the impellers of the ECP [21]

Materials Research Forum LLC
https://doi.org/10.21741/9781644902356

In pumps with a "floating" assembly, the impellers are not fixed to the shaft and can move along the shaft between the two guides. When the pump is running, each impeller is supported by the lower disk on the annular shoulder of the guide device. To reduce the friction force, a support washer made of wear-resistant material is pressed into the lower disk of the impeller [20, 21].

In the case of a "compression" pump assembly, due to the precise adjustment of the height of the hubs of the impellers, their contact with each other is ensured. The rings are fixed on the shaft and transmit their axial force to the shaft. To perceive such an axial load, a reinforced axial support is required [21].

Advantages of "compression" assembly:

- application in conditions of high content of mechanical impurities in the extracted products;
- unloading of impellers disks from axial load.

The disadvantage of the "compression" assembly is the complexity of installation. Since in order for the impellers not to come into contact with the guiding devices, when installing the pump, the gap in the working stages is adjusted by installing special calibrated plates between the shafts in the spline couplings [21].

In pumps made by "batch" assembly, several impellers and guide devices (from 3 to 10 pairs) are assembled into packages, while the height of the hubs of the impellers is selected in such a way that there is a small gap between the wheels. The peculiarity of such an assembly is the change of operating mode as wear and tear. As the support washers decrease, the distance between the wheels decreases, and the pump design becomes similar to the type with a compression assembly [21].

Downhole centrifugal pumps are multistage machines. This property distributes the value of the total pressure to all the steps involved. In turn, to achieve the required pressure, several pumping units are assembled sequentially, the number of which can reach 450 stages. Each stage can be replaced or removed from the chain if necessary [21].

1.3 Advantages and disadvantages of electric centrifugal pumps

Due to the minimum requirements for equipment at the mouth, ECPs may be in demand for applications on sites with limited working areas, such as offshore installations, if lifting costs are not a limiting factor. They are also used in fields where there is no available gas for gas lift systems. ECPs are one of the most high-volume methods of mechanized operation. ECPs have an advantage over other high-volume methods, since they can create a higher depression on the reservoir and increase its productivity in cases where it is possible to solve problems with interference from gas and sand removal. The diameter of the casing string is also not important to ensure the possibility of pumping such large volumes [22, 23].

As flooding increases, pumping out several thousand barrels of liquid per day in the process of improving the efficiency of reservoir displacement becomes traditional. This system can easily be automated and can be pumped periodically or continuously, but constant pumping is preferable to increase the service life, since the number of starts of the pump motor is limited to 100 – 150 times. The limited number of starts is due to high loads on the mechanical parts of the centrifugal pump (bearings, supports). For shallow wells, due to the relative ease of maintenance, capital costs are relatively low [22, 23].

The main problem of the ECP is a limited service life with a high content of mechanical impurities in the pumped liquid. The pump as such belongs to the high-speed centrifugal type, which can be damaged by abrasive materials, solid phase or debris. Critical wear of moving parts increases the risk of pump jamming. The formation of scale or mineral sediment may interfere with the operation of the centrifugal pump. The economic efficiency of the ECP largely depends on the availability of electricity. The disadvantage is manifested in remote regions, where the delivery of electricity is associated with large monetary costs. The lack of operational flexibility implies that all the main working components of the installation are located in the well at the level of the oil reservoir, and therefore, when a problem arises or a component needs to be replaced, the entire system has to be removed [22, 23].

If there is a problem of the possible presence of a high percentage of gas in the liquid (more than 30%), measures are taken to separate it and return it back to the casing before it enters the pump. The suction of large volumes of free gas can cause unstable operation and lead to mechanical wear and possible overheating of the electric motor with subsequent failure. In offshore installations where the use of a packer is required, all gas is pumped out with liquid. In these specific conditions, special pumps are used, in which it is possible to create a primary pressure at the pump intake [22, 23]. The primary pressure is created by injection machines similar to a centrifugal pump, but pumping the injection fluid or gas into the reservoir. Also, with a decrease in the efficiency of production by an electric centrifugal pump, as the reservoir is depleted, injection wells maintain the nominal pressure value at the pump inlet.

The solution to the problem or mitigation of the consequences of the presence of gas in the centrifugal pump is to develop a cooling system of the ECP to neutralize overheating. The latter measure will reduce the impact of the harmful effects of idling on a high-speed engine. Also, the adjustment of the protective equipment of the constant electronics in the ground control part will perform the functions of controlling the speed of the centrifugal pump motor from the load on the turbine.

It is possible to use an emergency braking mechanism based on the same principle of constant electronics, but with a sharp decrease in the operating current on the engine, the shaft is completely blocked with subsequent pumping or flushing of the well. The conclusion of the main causes of failures will clarify the need for the use of modification design or technological measures.

The diagram (Figure 6) shows the causes of premature failures of ECP installations and their share in the total number of failures.

It can be seen from the diagram that most cases of plant failures occur due to clogging with mechanical impurities. Mechanical impurities, interacting with turbines at high speeds, destroy the surface of the parts, as a result of which the hydrodynamic parameters of the stage deteriorate at the site of damage. The causes of mechanical damage to the cable, poor-quality preparation of wells, etc. are associated with improper operation of the pumping unit or due to errors made during commissioning [24 – 27].

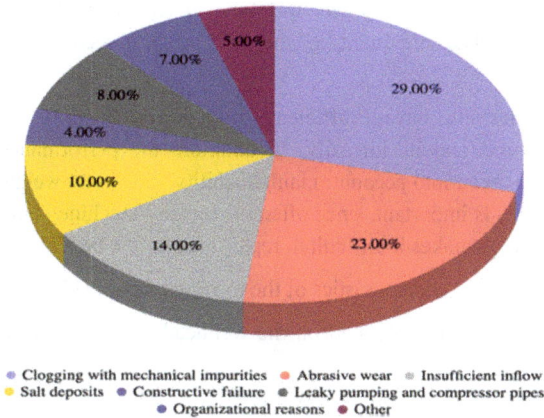

Figure 6. Diagram of the causes of premature failures of the ECP [24]

1.4   Rod depth pumping plant

The plant of a rod pump (Figure 7) refers to a volumetric type of pump, the operation of which is provided through a column of rods by the reciprocating working movement of the rocking machine. The uppermost rod is called a polished rod, it passes through the stuffing box of the wellhead fittings and connects to the head of the rocking machine balancer using a traverse and a flexible cable suspension of the column of rods.

When the plunger is moved upwards, due to an increase in volume, a vacuum is created in the cylinder chamber, as a result of which the liquid enters the pump cavity through the suction valve. The shut-off element of the discharge valve is pressed against the seat at this moment, which means that this valve is closed. During the movement of the plunger down, the volume of the working chamber decreases and the pressure in it increases. Under the influence of this pressure, the suction valve closes and the discharge valve opens. The liquid begins to flow through the pressure valve into the cavity above the plunger. At each cycle, a new portion of liquid will flow into the well cavity, which will gradually rise up the well [28].

Each rocking machine has individual parameters that depend on the required operational properties. However, along with them, this type of equipment has common technical characteristics. To analyze the quality of the machine, it is recommended to familiarize yourself with them.

All rocking machines must have a sufficiently high performance. It is determined by the movement of the rod and its intensity. In addition, the performance of the rocking machine should be taken into account: maintainability, size, total weight and complexity of maintenance. This is important, since often the rocking machine is installed away from populated areas, which makes it difficult to repair in case of a breakdown [29].

List of the main technical characteristics of the rocking machine [29]:

- the maximum permissible load on the wellhead rod, which can vary from 30 to 100 kN;
- the stroke length of the rod, usually it is from 1.2 to 3 m;
- the torque of the output gearbox shaft, which affects the intensity of the rod movement and can be from 6.3 to 56 kNm;
- the number of moves of the balancer varies from 1.2 to 15 per minute.

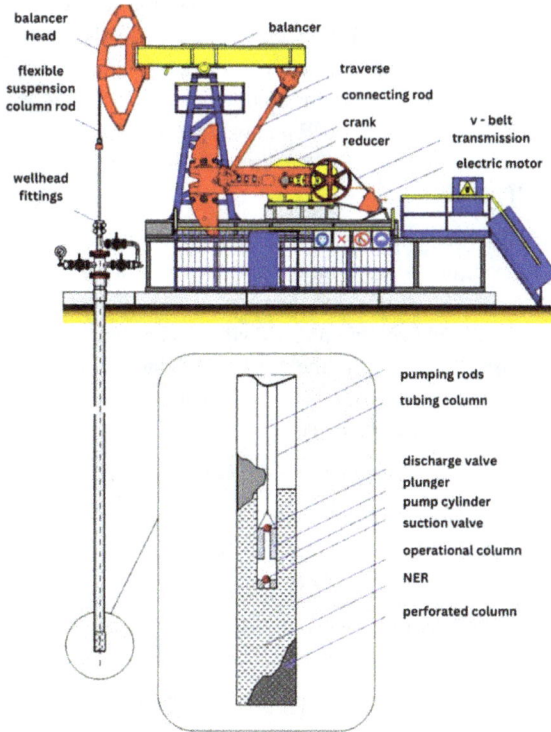

Figure 7. Unit of a downhole rod depth pump [30]

1.5   Design features of the rod depth pump

According to GOST 31835-2012 [31], the designs of rod pumps are divided into two main types – plug-in and non-plug-in. The non-plug-in pump (NP) is made in the form of a PCP pipe and uses it as a cylinder in which the plunger moves. The non-removable pump replaces the PCP pipe section and descends with it as the column is installed. The plug-in borehole pump (PP) is installed in the PCP after the tubing is installed in the well and can be removed at any time without the need to disassemble the column, but requires a larger PCP diameter.

Plug-in RDP are also divided into two types of devices:

-   plug-in pumps with top lock arrangement (PP1);
-   pumps, the lock of which is located in their lower part (PP2).

A lock is a device that serves to attach a plug-in pump in a PCP string. Locks with an upper fastening are used more often in the design, since the performance indicators indicate that this type of lock has a lower risk of pump failure compared to the lower fastening of the lock. Plug-in devices are used mainly for servicing wells of great depth up to 3,500 meters. The use of such RDP pumps, for the extraction of which it is enough to lift the rods with which the entire pump structure is connected, greatly simplifies the repair of the well, if necessary.

Extraction of the NP type pump is carried out in two stages: first of all, a plunger with valves is removed from the pump cylinder, and then a cylinder with PCP is lifted from the well. Due to the increased working diameter of the cylinder, relative to the plug-in pumps, the mass of the liquid column presses with greater force on the valve of the pump plunger. In this regard, an increase in the depth of production over 1,500 meters sharply increases the risk of breaking the rods. NP is advisable in wells with a large flow rate and a small depth of descent up to 1,500 meters [31].

Non-plug-in pumps are also divided into several categories [31]:

 – pumping units without a catcher (NP);
 – non-plug-in depth pumps with gripping rod (NP1);
 – non-plug-in pumps with a catcher (NP2).

The technical characteristics of standard pumps in accordance with GOST 31835-2012 are presented in Table 1.

Table 1. Specifications for standard pumps [31]

Pump type	Designation	Pressure, [m]	PCP diameter, [mm]	Plunger working stroke, [mm]
Plug-in with top lock	PP1B-27	2,500	60	
	PP1B-32	2,200		
	PP1B-38	2,000	73	
	PP1B-44	2,000		
	PP1B-57	1,500	89	
Plug-in with bottom lock	PP2B-27	3,500	60	
	PP2B-32	3,500		
	PP2B-38	3,500	73	1,200 – 4,000
	PP2B-44	3,000		
	PP2B-57	2,500	89	
Tube pumps with a catcher	NP2B-57	1,500	73	
	NP2B-70	1,200	89	
	NP2B-95	1,000	114	
Tube pumps without a catcher	NPB-57	1,500	73	
	NPB-70	1,200	89	
	NPB-95	1,000	114	

The following diagram (Figure 8) is a diagram of the dependence of the diameter of the working part of the cylinder and the depth of the pump extraction. The absence of a difference between pipe pumps is due to the same cylinder diameters.

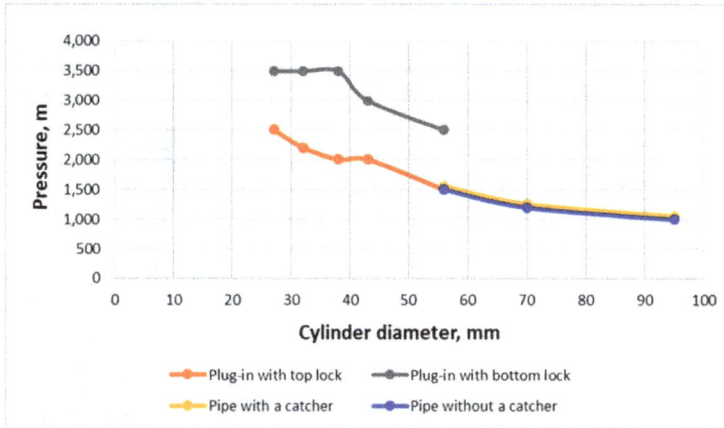

Figure 8. Pressure dependence on pump type

Downhole pumps of the PP1 version (Figure 9) are designed for pumping low-viscosity fluid from oil wells. High efficiency of low-viscosity liquid production can be achieved by using two sequential valve pairs in the pump design. The need for additional valves is due to the increased fluidity of the oil-containing fluid due to the low viscosity of the produced fluid. The pump consists of a one-piece cylinder (6), on the lower end of which a double suction valve (9) is screwed, and on the upper end there is a lock (1) of the plunger, movably located inside the plunger (7), on the threaded ends of which are screwed: from below a double discharge valve (8), and on top is the plunger cage. To connect the plunger to the pump rod string, the pump is equipped with a rod (2) screwed onto the plunger cage and secured with a lock nut (4). In the bore of the upper sub of the cylinder there is a stop (3), resting against which, the plunger ensures the failure of the downhole pump from the lock support [31].

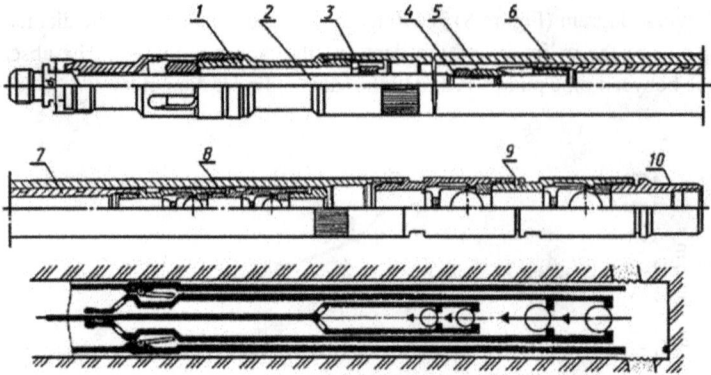

Figure 9. Pump diagram PP1 [31]: 1 – pump lock and seal; 2 – stem; 3 – stop; 4 – locknut, 5 – plunger sub; 6 – cylinder; 7 – plunger; 8 – discharge valve; 9 – suction valve; 10 – pump sub

Downhole pumps of the PP2 version (Figure 10) have a similar design to the pump of the PP1 version, but due to the design features of the lower lock fastening, it is used in a wider range of working depths up to 3,500 meters.

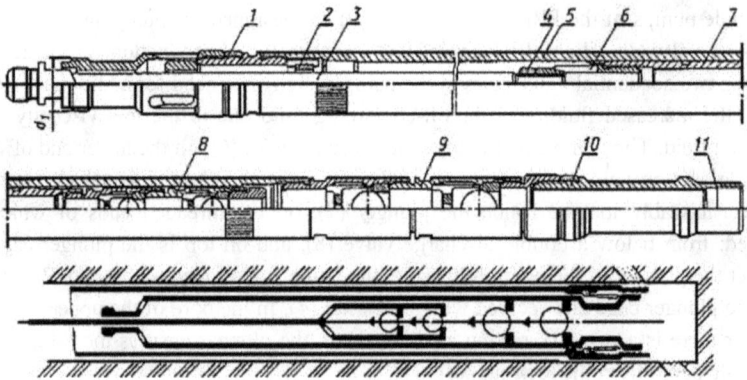

Figure 10. Scheme of the PP2 pump [31]: 1 – protective valve; 2 – emphasis; 3 – stock; 4 – locknut; 5 – cylinder; 6 – plunger sub; 7 – plunger; 8 – castle support; 9 – suction valve; 10 – persistent nipple with a cone; 11 – pump sub

The increase in the installation depth is due to the great influence of the downforce of the oil column on the pump housing, which in turn improves the sealing properties between the rings located between the lock support on the PCP and the pump lock. The mating parts of the sealing rings are made in the form of polished inner and outer cones, and when the pump is installed tightly, with a tight fit, they are pressed against each other, and thereby seal the area above the pump and the area under the pump from liquid leakage [31].

The NP1 pump (Figure 11) consists of a cylinder (1), plunger (5), discharge and suction valves. In the upper part of the plunger there is a discharge valve (3) and a stem (2) with a sub (9) for the rods. Suction valve (7) is freely suspended from the upper end of the plunger by means of the tip (6) on the gripping rod (4). During operation, the valve sits in the body seat (8).

Hanging the suction valve from the plunger is necessary to drain the fluid from the PCP before lifting them, as well as to replace the valve without lifting the PCP. The presence of the capture of the rod inside the plunger limits its pressure, which in the NP1 pumps does not exceed 0.9 m. In this regard, the NP1 pump is effective when the large cylinder diameter exceeds 57 mm to 95 mm [31].

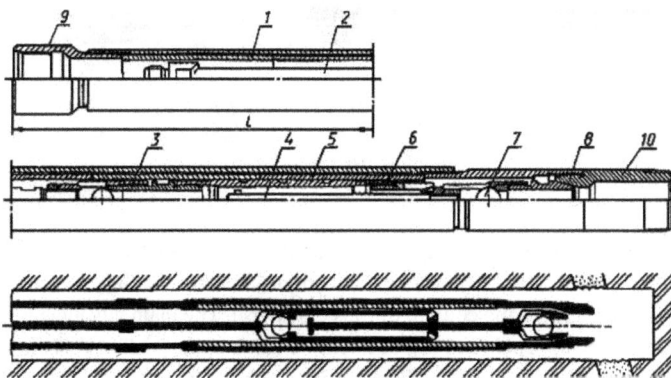

Figure 11. Scheme of the NP1 pump [31]: 1 – cylinder; 2 – stock; 3 – discharge valve; 4 – gripping rod; 5 – plunger; 6 – plunger tip; 7 – suction valve; 8 – cone seat, suction valve support; 9 – top sub; 10 – bottom sub

In the pump NP2 (Figure 12), unlike the pump NP1, the discharge valve (5) is installed at the lower end of the plunger (4). To extract the suction valve (7) without lifting the PCP, a catcher (6) is used in the form of a bayonet lock, which is attached to the seat of the discharge valve.

The catcher has a cylindrical shape with two grooves in the form of hooks for the hook, the grooves are made mirror-like and allow you to hook when turning the catcher clockwise. A shortened rod is screwed into the suction valve cage with installed studs for hooking. After installing the suction valve in the seat of the housing (8) and securing it to the seat, the column of rods rotates counterclockwise and, together with the plunger, is removed from the hook zone.

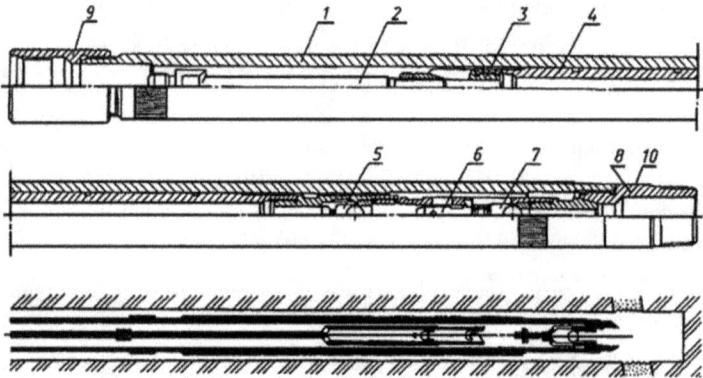

Figure 12. Scheme of the NP2 pump [31]: 1 – cylinder; 2 – stock; 3 – plunger sub; 4 – plunger; 5 – discharge valve; 6 – catcher rod; 7 – suction valve; 8 – body saddle; 9 – top sub; 10 – bottom sub

The capture and extraction is not carried out until it is necessary to replace or repair the working parts of the pump. The capture is carried out after landing the plunger on the shortened rod and turning the column of rods clockwise. The main advantages of the NP2 design are the ease of repair and replacement of valve pairs "saddle-valve". The design uses only one technologically difficult part in production – a cylinder [31].

1.6 Main details of the rod depth pump

The valves of deep well pumps are made with ball valves, since under the operating conditions of deep well pumps they have the property of uniform wear during suction cycles. Compared to other (tapered and flat) designs, the ball valve has the ability to turn or rotate, allowing the entire surface to be used and supported. The long service life, which exceeds the service life of the entire pump, is due to the high-quality grinding of the ball to the seat during operation, while maintaining the size of the ball for a long time due to the use of the entire working surface. The absence of sharp corners as places of concentration of deposits prevents the formation of harmful sticking on the surface of the valve, as well as places of concentration of corrosion damage.

Depending on the design of the seat, ball valves (Figure 13) are available with a shoulder and with a smooth outer surface. The latter are used, as a rule, as discharge valves due to their smaller dimensions and ease of attachment.

a)   *b)*

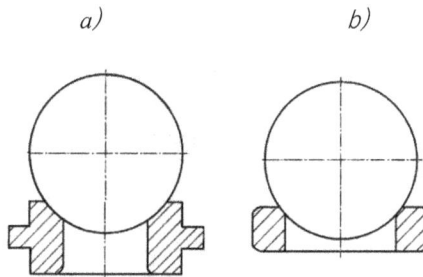

Figure 13. Ball valves with collar (a) and smooth outer surface (b) [31]

The valve seats are symmetrical and when one of the edges of the seat surface is worn, they are rotated (rearranged) by 180° to use the other surface. To ensure the tightness of the "ball-seat" joint, the inner edge of the seat has a 15° chamfer with a transition to R1-2.5 mm rounding. The hardness of the ball is always set higher than the hardness of the seat, since the ball must retain its shape during operation. Ball hardness is usually 56...70 HRC, saddle 40...50 HRC. The ball and seat are made of high carbon steel, hard alloys, and in some cases (for example, in a corrosive environment) of bronze or ceramic. The most commonly used saddles and balls are made of hard alloy Stellite, since the alloy combines the properties of high hardness, strength and relatively easy processing, as well as the best price-quality ratio – up to 50 US dollars.

The development of technology for the production of modern magnets has made it possible to use them in the design of the valve. This solution is described in the patent "Self-aligning magnetic valve of the rod depth pump" (RF Patent, RU 185543 U1, IPC F L 51/00 [32]), the utility model of which relates to the oil industry, in particular to sucker rod pump units and can be used to operate oil wells, including horizontal and obliquely directed. A self aligning magnetic valve of a sucker rod pump, containing a housing in which a valve pair is placed, consisting of a ball interacting with a metal seat, under which a magnetic ring is installed, having dimensions similar to the seat.

The plunger of the deep well pump (Figure 14) is a steel pipe with an internal thread at the ends. The length of the plunger varies (from 500 mm to 1,800 mm) and depends on the required stroke of the plunger and the length of the pump cylinder. They are made of Steel 45, 40Cr or 38CrMoAlA. According to the method of sealing the "cylinder – plunger" gap, completely metal and gummed plungers are distinguished. In a pair of metal plunger – cylinder", the seal is created by a normalized gap of great length, in gummed ones – due to cuffs or rings made of elastomer or plastic. The types of plungers are divided into: smooth plungers, plungers with annular grooves, plungers with a cylindrical bore and a beveled end, a cuff or gummed plunger [31].

The use of a large number of different plunger designs is due to the need to ensure the tightness of the gap under any operating conditions, high durability of the "cylinder – plunger" pair, while striving to reduce the friction forces as much as possible.

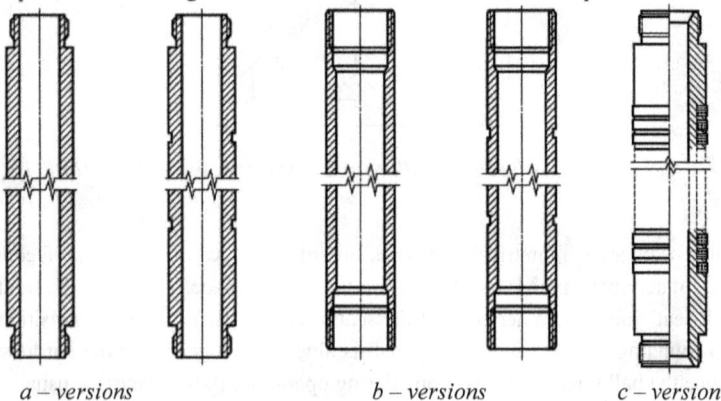

a – versions   b – versions   c – version

Figure 14. Plungers [31]: a – smooth and with annular grooves with external thread; b – smooth and with annular grooves with internal thread; c – cuff, gummed plunger

To ensure high durability of the pump, it is of great importance to prevent scuffing of rubbing surfaces. The reason for this phenomenon is both the abrasive contained in the pumped liquid and the appearance of local dry friction zones of the "plunger – cylinder" pair as a result of a rupture in the gap of the pumped liquid film.

To ensure the normal operation of a pair of mated parts, plungers with recesses and grooves are used, or the hardness of the working surface of the plunger is increased by cementation, spraying of carbide powders, nitriding or chrome plating. Chrome-plated plungers are the most durable and have a lower coefficient of friction than cemented ones. In addition, the chromium layer provides good corrosion resistance when working in wells with a high SO_2 content. It should be noted that chrome plating is a relatively expensive process, as a result of which plungers are not chrome-plated, but made of carbon steel, hardened with high frequency currents, are more widely used. Also, the modern solution to this problem is the coating of the cylinder and plunger with antifriction and hydrophobic coatings.

According to the size of the gap between the cylinder and the plunger, pumps are divided into three groups [31]:

- group I (tight fit of the plunger) with a gap between the plunger and the cylinder from 0.02 to 0.07 mm, is designed for lifting low-viscosity reservoir fluid with a low sand content, increased gas release at great depths of the pump suspension;
- group II (medium fit) with a gap from 0.07 to 0.12 mm, designed for lifting medium viscosity reservoir fluid with a high gas content at medium suspension depths;
- group III (weak landing) with a gap of more than 0.12 mm, is designed for lifting very viscous oil from heavily watered wells at a shallow depth of the pump suspension.

Cylinders of rod pumps are produced in two versions [31]:

- CW – one-piece (without bushing), thick-walled;
- CC – compound (bushing).

The cylinder of the bush pump consists of a casing in which bushings are placed. The bushings are fixed in the casing with nuts. The bushings are subjected to a variable internal hydraulic pressure caused by the pumped liquid column and a constant force resulting from the end compression of the working bushings. Bushings of all compound cylinders with different internal diameters have the same length up to 300 mm.

The bushings of all pumps are made of three types: alloy steel grades 38CrMoAlA, steel from grades 45 and 40Cr, cast iron grades GI26-48. Alloy bushings are made only thin–walled, steel - thin-walled, with increased wall thickness and thick–walled, cast iron - thick-walled. To increase durability, the inner surface of the bushings is strengthened by physico-thermal methods: cast iron – hardened with high frequency currents, steel – nitrided or cemented [31]. As a result of this treatment, the hardness of the surface layer is up to 70 HRC.

Mechanical processing of bushings is carried out by grinding and honing. The main requirements for the machining of bushings are a high class of accuracy and cleanliness of the inner surface, as well as the perpendicular ends to the axis of the bushings.

Deviations from the geometry of the inner diameter of the cylinder bushings remain within the limits of no more than 0.03 mm. Thus, the manufacture of a sleeve cylinder requires compliance with a variety of accuracy parameters and imposes requirements on the accuracy of the equipment, but when used in a pump, it allows repair of parts with a worn inner surface.

Solid cylinders are a long steel pipe, the inner surface of which is working. At the same time, the pipe performs the functions of both the cylinder and the casing at the same time. Such a design is devoid of such disadvantages as leakiness between the ends of the working bushings, curvature of the cylinder axis. At the same time, the rigidity of the pump increases and it becomes possible to use a large-diameter plunger with the same outer diameter compared to the bushing cylinder [31].

1.7   Patents of non-standard pump designs

The designs of rod oil pumps are not limited to the types listed in GOST 31835-2012 [31]. Heterogeneous conditions that negatively affect the operation of pumping units have influenced the emergence of a wide range of design solutions to neutralize their negative impact.

Known downhole rod pump (RF Patent, RU 2321772 C1, IPC F04B 47/02 [33]) for the extraction of high-viscosity oil containing a fixed cylinder, a movable cylinder and a fixed plunger. Inside the movable cylinder, a stationary hollow plunger is installed, connected in the lower part to a hollow receiving pipe, made with an inner diameter equal to the inner diameter of the plunger and an outer diameter smaller than the outer diameter of the plunger. The design of the pump provides increased efficiency by increasing its performance and expanding the range of use of the pump. The disadvantages of this design are the gap between the movable cylinder and the fixed plunger and the intake pipe. Mechanical impurities penetrating into the gap increase the risk of pump jamming, broken rods, water hammer, resulting in damage to the cylinder.

A pump design is known (RF Patent, RU 96617 U1, IPC F04B 47/00 [34]) containing a column of rods, upper and lower cylinders with upper and lower plungers connected to each other by a rod, upper and lower controlled valves, characterized in that the plungers contain spring-loaded ball valves, and the upper cylinder is mounted in a lock support, and the lower cylinder is fixed. This design combines two types of pump. The upper part is made according to the type of plug-in pump with a lock support, and the lower part repeats the features of a non-plug-in pump. Controlled valves with springs allow to minimize hydrodynamic losses when the pump is closed. The disadvantage of this design is the use of springs, jamming of which is possible with a high content of mechanical impurities and ARPD. This phenomenon occurs as a result of a large number of sediment concentrators in the form of a spring. Another disadvantage is the reduction of the useful volume due to the embedded elements of controlled valves.

The plug-in rod pump (RF Patent, RU 163755 U1, IPC F04B 47/02 [35]) contains in its design one movable plunger connected to a column of rods and fixed, which is fixed inside the PCP column in a lock support. The plungers are connected by a movable cylinder. As a result of lifting the cylinder, the useful space between the plungers increases. The advantages of this design are the shorter length of the cylinder, which facilitates their production and with smaller pump sizes (in the shifted state) during operation has a working volume equal to comparable with standard pumps of large sizes. Jamming of the pump, as a result of stopping the pump and deposition of mechanical impurities, since there is a gap between the PCP and the movable cylinder and, therefore, the loads during pump start-up decrease.

A "Rod pumping unit" is known (RF Patent, RU 2175402 C1, IPC F04B 47/00 [36]), containing columns of pumping pipes and rods, a cylinder with steps of different diameters installed one above the other, two hollow plungers with suction and discharge valves connected to each other and with a load placed outside the PCP cavity, movably located in the cylinder to form working chambers of smaller and larger diameters.

The disadvantages of the prototype include the fact that a "saddle-ball" pair is used in the suction valve, which significantly limits the flow sections of the valve assembly, which means additional hydraulic resistances in the valve assembly. In addition, the set of loads increases the dimensions of the pump, to some extent complicating the installation and disassembly of the pump.

Examining the shortcomings of pumps of non-standard designs, it can be concluded that the vast majority of the causes of pump malfunctions occur as a result of the failure of the valve pairs "saddle-ball" and mechanical wear of the rubbing parts of the cylinder and plunger. Also, the non-standard pump designs presented in the patents imply the use of additional difficult-to-process parts requiring high-precision machining on the inner and outer surfaces of additional plungers, as well as springs and valves of a new design.

1.8   Causes of malfunctions of rod depth pumps

One of the most important factors influencing the failures of rod pumping units are the properties of the extracted liquid. It is known that mechanical impurities contained in the pumped liquid lead not only to abrasive wear of the pump and equipment, but also to complex accidents. If the downhole fluid is pumped out before receiving the pump, the rocking machine will automatically turn off to accumulate fluid in the well. When the pump stops, mechanical impurities are deposited above the pump (forming a column up to 20 m in height), fall into the gap between the plunger and the cylinder, which causes the plunger to jam. Attempts to start the pump after stopping by running the plunger in most cases lead to the appearance of scuffles on rubbing surfaces, an increase in the gap of the cylinder-plunger pair, an increase in leaks, a decrease in the feed ratio or an accident (rupture of rods) [37 – 39]. To prevent the breakage of rods and abrasion of pipes, hardened surfaces of couplings with oval edges and treated with high frequency currents, as well as couplings with spraying on the outer surface of a hard alloy, are used.

Mechanical impurities enter the pump from the production column. Solid precipitation falls first into a protective device (filter) mounted at the reception of the rod equipment, then with a reduced concentration of mechanical impurities enter the rod pump and significantly affect the operability of the plunger and valve pairs of equipment. To combat paraffin deposits, periodic thermal treatments of wells are also carried out, without stopping them, by pumping hot oil into the annulus, which, passing through valves and PCP, melts paraffin deposits and brings them to the surface. Glazed or coated with a special varnish PCP is also used, on which paraffin does not settle [40 – 43].

The ingress of sand negatively affects the threaded connections of pumping pipes at the slightest leakiness of the connections, especially in waterlogged wells, it quickly corrodes the thread and liquid flows through the formed channel, which prevents the establishment of the optimal mode of operation of the well. The presence of a large amount of poorly permeable sediments at the bottom of the well primarily leads to a decrease in the flow rate of the liquid, because the concentrated mixture in the well increases the back pressure on the bottom and worsens the conditions of natural fluid inflow. Technical or technological shutdowns of wells contribute to the deposition of sand on the bottom and the formation of traffic jams, which is often the most serious problem in the operation of sand wells. At the same time, it should be borne in mind that the main amount of formed plugs is obtained precisely as a result of sedimentation from a column of liquid in the column [41, 44].

In order to prevent the settling of mechanical impurities in the area of the bottom-hole zone, it is necessary to maintain the condition of constant pumping of liquid until the concentration of sand in the extracted products decreases.

The intensity of sand intake depends on a complex of factors [41, 44].

- filtration rate, or the magnitude of depression;
- the degree of cementation of the soil of the geological rock;
- viscosity and density of well production;
- and other physico-chemical properties of the liquid.

At the same time, it must be assumed that all factors except the filtration rate or pressure remain constant. The filtration rate or pressure gradient depends on the amount of fluid extraction from the well. It follows that by changing the number of double strokes of the rocking machine, it is possible to influence the sand appearance [44].

Thus, it is possible to identify the main measures to combat sand at the pump intake:

- unit of filtering devices (sand anchor, gas-sand anchor and filter) for filtration at the pump intake;
- regulation of fluid extraction from the well.

If there are noticeable leaks in the pipes, cylinder and valve pair of the plunger, then the liquid level in the annular space will increase not only due to the inflow of liquid from the reservoir, but also due to the flow of liquid from the transmission elements of underground equipment. At the same time, the liquid level in the PCP will decrease, that is the pumping-compressor pipes will be partially emptied.

If the flow rate of fluid from the pumping-compressor pipes and other transmission elements is equal to the amount of fluid flowing from the reservoir (per unit of time), in this case, during the shutdown of the pumping machine, the volume of fluid accumulated in the inter-tube space of the well will partially consist of fluid flowing out of the pipes and other transmission elements. It follows that the flow rate of the well during operation will significantly decrease compared to the expected. An increase in the volume of leaks during the operation of the pump will affect not only the flow rate of the well for oil, but also the speed of the upward flow of liquid, the speed will decrease.

During the operation of downhole rod equipment, leaks in the valve pair of the pump and in the gap between the plunger and the cylinder inevitably lead to losses in fluid production, even if leaks that have arisen in the pumping-compressor pipes are not taken into account. At the same time, it should be borne in mind that the presence of leaks during continuous pumping of liquid may not directly affect the flow rate of the well, leading only to a reduction in the inter-repair period. With periodic pumping, more or less significant leaks directly affect the extraction of liquid and therefore the leak must be eliminated [41, 44].

To solve this problem, the following activities are carried out:

- − use pumps with reduced harmful space;
- − increase the stroke length of the plunger;
- − increase the depth of immersion of the pump under the liquid level in the well;
- − the gas is pumped out of the annular space.

According to the results of statistical studies of the causes of failures, the most likely factors from the above that lead to RDP failures were identified (Table 2).

The studied factors affecting the performance of the RDP, in most cases, lead to two failure variants with different degrees of repair complexity. For example, restoring the operation of the installation when the pump is jammed is not comparable in labor costs with replacing valve pairs or flushing the well, because in the first case, one of the likely solutions to the problem is lifting the PCP string.

Table 2. Results of the analysis of the reasons for failures [45]

Well number	Type of pump failure	OHP, [day]	Debit, [m³/day]	Water cut, [%]	Suspected cause of pump failure	Pump descent depth, [m]	Dynamic level, [m]
1	Breakage of rods	65	7.6	4	Mechanical impurities, ARPD	1,134	1,125
2	Failure in the operation of the RDP	175	5	2	Valve sticking	1,460	1,440
3	Failure in the operation of the RDP	352	2.2	1	Pump spell	1,156	1,085
4	Breakage of rods	758	12.2	3	Large operating time	1,224	1,031
5	Lapel rods	84	4.8	1	Mechanical impurities, ARPD	1,301	1,280
6	Breakage of rods	227	25.4	0.5	Mechanical impurities, ARPD	1,270	1,240
7	Failure in the operation of the RDP	150	4.7	0.1	Valve sticking	1,160	1,130
8	Breakage of rods	94	4.8	1	Mechanical impurities, ARPD	1,493	1,460

Based on the data obtained, a distribution diagram of the main causes of failures of a rod pumping unit is constructed (Figure 15), which shows that one of the main factors having a significant impact on RDP failures is the presence of mechanical impurities in oil (Figure 16, 17).

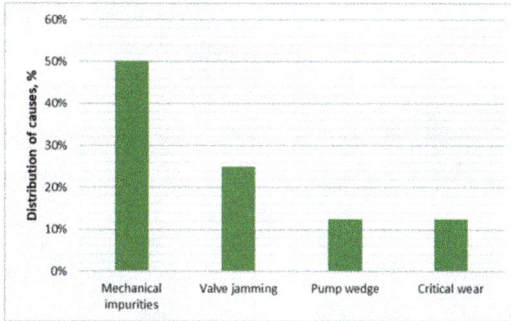

Figure 15. Diagram of the distribution of the main causes of failures

More than half of all failures occurred as a result of the breakage of the pumping rods (Figure 16).

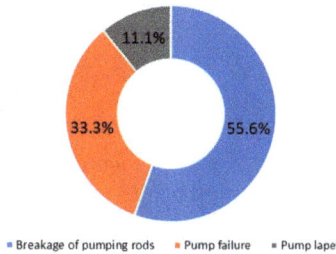

Figure 16. Diagram of the consequences of failure of the pumping unit

The reasons for the breakage and failure were:

   – mechanical impurities (Figure 17, 18) that caused the pump to jam;
   – jamming of the valve as a result of ingress of high-viscosity inclusions, asphalt-resin-paraffin deposits (ARPD) and salt (Figure 19).

It should also be noted that other properties of the extracted liquid have a significant impact on the performance of the RDP, namely:

- the amount of free gas in the operational flow;
- work in an aggressive environment;
- deposits of salts and ARPD on the working nodes of the RDP (Figure 19, 20);
- high viscosity of the produced reservoir fluid.

Figure 17. Choking with mechanical impurities of the RDP plunger

Figure 18. Abrasion and breakage of threaded PCP connections

Figure 19. Deposits of gray solids (salts) on the inner surface of the suction valve seat

Figure 20. Paraffin deposits on the rods

These factors lead to premature failure of wells equipped with rod depth pumps, clogging the working elements inside the downhole equipment.

It is known that mechanical impurities, such as paraffin, salts contained in the pumped liquid, lead not only to abrasive wear of the coupled elements of the pump, but also to the formation of neoplasms and cementation of the moving elements of precision pairs [46 – 48]. The formation of a foreign medium in the coupled parts of the pump leads to a violation of the design trajectory of the movement of equipment elements, which leads to jamming of the main units and units of pumping equipment.

Materials Research Forum LLC
https://doi.org/10.21741/9781644902356

With an increase in the speed of oil movement, the intensity of deposits initially increases, which is explained by an increase in turbulence of the flow and, consequently, an increase in the frequency of formation and separation of bubbles from the pipe surface, floating suspended particles of paraffin and asphalt-resin substances. The roughness of the walls and the presence of solid impurities in the system also contribute to the release of paraffin from the oil into the solid phase. Therefore, the choice of optimal ways to combat asphalt-resin-paraffin deposits is an urgent and important direction in the domestic oil industry.

1.9 Methodology for calculating the dosage of salt deposition inhibitors

The task of reducing ARPD in modern practice is solved by introducing inhibitors of the most common class, however, the desired result has not been achieved due to the lack of a methodology for their qualitative assessment and dosage [43, 49 – 59]. In this connection, studies on the chemical and physical composition of the oil liquid were carried out at the domestic wells of the Uzen field. To identify the factors affecting the operation of RDP, basic studies of the compositions of 8 oils, 1 ARPD and 11 reservoir waters were carried out [60 – 64].

According to the results of the analysis of oil compositions, 6 paraffin deposition inhibitors were reasonably recommended to study the effect of chemical reagents on these oils: SNPH-7909, SNPH-7941, SNPH-7963, FLEC IP-106, RT-1M and RTF-1 (Figure 21).

Due to the fact that oil under normal conditions (t = 20 °C) is a mushy mass and washing of the film by the Petrolight method is not possible, an analysis of the relative decrease in the surface tension of oil at the water-oil phase boundary with various inhibitors at 50 °C and 60 °C on the tensiometer was carried out. The conclusion was that the lower the surface tension index, the lower the ability of the oil fluid to form large chains of paraffin molecules, and the easier it is to carry them to the surface with the flow of liquid.

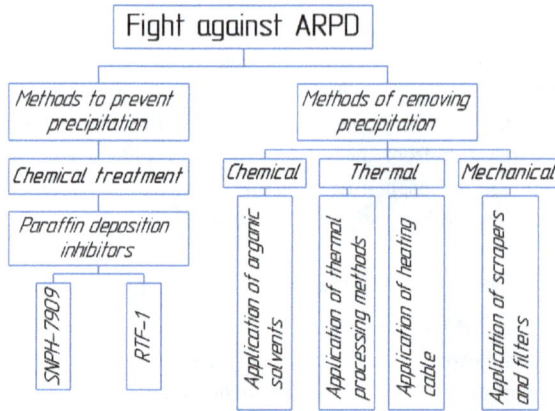

Figure 21. Ways to prevent and remove ARPD

A well with a low OHP (up to 60 days) was selected for deposits that occur during operation, namely: clogging of the pump cavity with salts and jamming of the RDP. The results of the investigation of the causes of premature failure of well No. 4406 of the Uzen field before the start of the OPI were obtained. Failures of underground equipment were recorded on 14.01.17, 17.02.17, 13.05.17 and 23.06.17. The operating time for failure was 53, 33, 85 and 40 days, respectively. The reason for the failures of underground equipment is the jamming of the RDP due to clogging with salts. On the rods (Figure 22, a), pumping-compressor pipes (Figure 22, b), plunger (Figure 22, c), deposits of gray solids (salts) were found.

According to the results of the inspection of the RDP in the workshop for the diagnosis of underground equipment, the receiving filter has deposits of mechanical impurities and salts, the suction valve in the salt deposits (Figure 23 a, b).

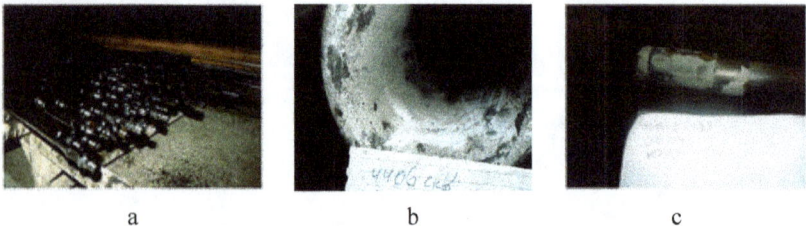

a   b   c

Figure 22. Deposits of gray solids (salts): a – rods; b – pumping and compressor pipes; c – plunger

a   b

Figure 23 – Suction valve in salt deposits: a – ulcerative corrosion of the metal surface of the seat and ball; b – salt growths along the inner diameter of the suction valve seat

The method of combating salt deposits is the periodic delivery of inhibitors [65, 66]. In this case, the component composition of the oil and the parameters of the equipment indicate that the inhibitor should be fed manually through the annulus.

The inhibitor is a brown water-organic solution designed to prevent deposits of inorganic salts in downhole and underground equipment during oil production, transportation and preparation. Table 3 presents the main quality indicators and metrological support for determining the compliance of these indicators with the requirements of regulatory and technical documentation. It has been experimentally established that the reagent should be dosed at a concentration of 100 g/m^3 in the form of an inhibitor solution at the rate of 30 liters of solution per day. The method of calculating the dosage of the inhibitor was developed according to the actual parameters of the Q_{liquid} and Q_{water} during the operation of the well. The daily volume of the inhibitor is calculated according to the formula (1):

$$V_{chem.\,reagent} = \frac{Q_{water} \times 100}{1,000 \times 1.06}, \qquad (1)$$

where $V_{chem.\,reagent}$ – daily amount of inhibitor, dm^3; Q_{water} – daily volume of extracted associated water per day, m^3; 100 – dosage, g/m^3; 1,000 – conversion factor of the amount of inhibitor in kg; 1.06 – inhibitor density, kg/dm^3.

After applying the inhibitor, it was found that no salt deposits were found on the outer and inner parts of the suction valve body, the surface was clean (Figure 24, a and 24, b). The plunger was without mechanical damage, the surface was clean, without scratches (Figure 24, c).

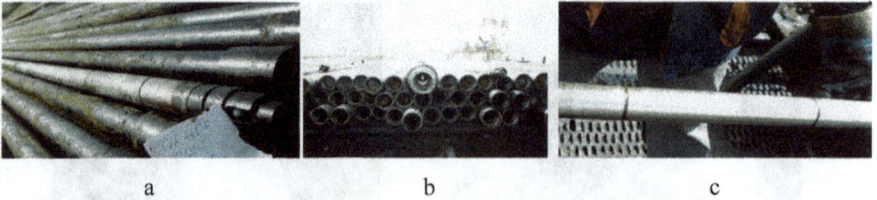

Figure 24. Underground equipment for mechanized oil production: a – the outer part of the suction valve body; b – the inner part of the suction valve body; c – RDP plunger

During the application of the inhibitor, failures were recorded on 21.08.2017 and 17.11.2017. The operating time was 56 and 86 days. The reason for the failures is the breakage of the column of pumping rods in the amount of 51 pieces from the wellhead. No salt deposits were found in the outer and inner parts of the suction valve body (Figure 25, a). The discharge valve is clean, sealed during crimping (Figure 25, b). The plunger is free of mechanical damage and scratches, scratches, the stroke is free without wedging. The pump is suitable for further operation.

Figure 25. Deep-pumping equipment without salt deposits: a – suction valve body; b – seat and ball of the discharge valve

Based on the results of operation at the well and bench tests, a method for calculating the OHP of RDP is proposed. Before the OPI the OHP is calculated according to the formula (2):

$$OHP = \frac{\sum days\ worked}{N}, \qquad (2)$$

where OHP – overhaul period, day; $\sum$ days worked – the amount of days worked since the beginning of the year; N – number of RDP failures.

$$OHP = \frac{53 + 33 + 85 + 40}{4} = 53 \quad days.$$

During the OPI, the OHP is calculated according to the formula (2):

$$OHP = \frac{56 + 86 + 34}{2} = 88 \quad days.$$

According to the results of the OPI, Table 3 presents the comparative characteristics of the operation of the well before and after the use of the inhibitor.

Table 3. Summary of the results of studies on salt deposition inhibition

№	Indicators	Well No. 4406	
		Before the start of the OPI	Based on the results of the OPI
1	Q liquid, [m³/s]	47.5	48.2
2	Q oil, [m³/s]	5.6	6.2
3	OHP, [day]	53	88
4	Number of repairs	4	2
5	Reasons for failures	Jamming of RDP as a result of clogging with salts	The lapel of the safety valve and the breakage of the column of rods
6	Dosage of the salt inhibitor, [liters]	0	30 (every day)
7	Number of rejected RDP	4	0

Summing up the results of the OPI of the inhibitor in order to increase the OHP, conclusions were drawn:

- OHP increased 1.6 times;
- there were no well failures due to salt clogging;
- there was a decrease in the number of underground well repairs by 2 times;
- reduction of the number of defects of deep-pumping equipment.

The next measure to reduce oil volume losses is recommended to use a demulsifier of the Randem 2201 brand (hereinafter referred to as a demulsifier) in order to separate water during the production preparation of commercial oil. The determination of the compatibility of the demulsifier used at the Uzen deposit with the inhibitor was carried out using the "BOTTLE TEST" method, the essence of which is visual observation of the delamination of the emulsion and water separation in glass vessels (bottles for settling).

Laboratory measurements were carried out on a sample of water-oil emulsion of the Uzen field with a total water content of 13.6%. The production preparation of oil is the demulsification of oil, which is the process of heating oil at a temperature of 60 °C and supplying a chemical reagent. As a result of the combined effect of temperature and chemical reagent, coalescence occurs, that is, intensive merging of water droplets into larger ones that can fall out and separate from oil under the influence of gravity. Figure 26 shows the results of a study of the effect of an inhibitor on the oil treatment process, and its compatibility with a demulsifier (DE*).

Sampling and acceptance of the finished product was carried out according to GOST 3885-2012 [31]. The volume of the average laboratory sample was 500 cm^3. The water content in oil is determined according to GOST 2477-2014 [67], the essence of which is to heat a sample with a solvent insoluble in water and measure the volume of condensed water. In this case, the "BOTTLE TEST" study was conducted on the automatic stability analyzer of multicomponent dispersed systems "MultiScan MS 20". The results of the study were processed using the «MSC 20» software. The "MultiScan MS 20" device is designed to quickly and accurately determine the stability of emulsions and aging of colloidal systems, as well as to determine the effectiveness of salt deposition inhibitors. First of all, it provides reliable characterization of time- and temperature-dependent properties of multiphase systems.

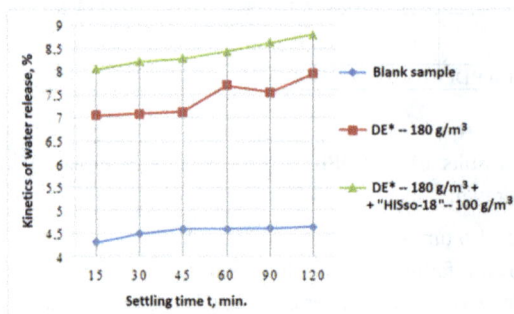

Figure 26. Kinetics of water release with the combined use of a demulsifier (180 g/m^3) and an inhibitor (100 g/m^3)

According to the results of laboratory studies, it can be seen that when an inhibitor and a demulsifier are introduced into the initial oil-water emulsion, there is no negative effect on the preparation process - dewatering of oil, where the volume of water released is 8.80%, and the residual water content in oil is 4.80%.

When a salt deposition inhibitor is used together with a demulsifier, the released water does not become cloudy, a clear phase boundary is observed, there is no intermediate layer, no sediment and adhesion to the walls of the test tube is formed.

The use of modern metrological support when changing the component composition of oil allowed us to conclude that the inhibitor is compatible with the demulsifier and contributes to an increase in the volume of released water.

The results of studies before and after the use of chemical reagents proved that it is necessary to recommend an inhibitor and a demulsifier for industrial use in order to increase the oil recovery of wells and the inter-repair resource of RDP during oil production.

It is proposed to solve the technical problem of increasing the overhaul period of the PRDP by eliminating the ARPD with the introduction of scientifically-based inhibitors into the system by the authors of the project (SNPH-7909, SNPH-7941, RTF-1). Therefore, the scientific and technological measures proposed by the authors of the study will significantly reduce resource, technological and financial costs, increase the productivity of rod depth pumps, the oil recovery coefficient.

1.10 Methods of restoring pump operability

Underground well repair is a complex of works related to the prevention and elimination of problems with underground equipment and the borehole. Underground well repair is conditionally divided into current and capital. Current repairs are divided into planned preventive (or preventive) and restorative [68].

Planned preventive well repair is a repair in order to prevent deviations from the specified technological modes of well operation caused by possible malfunctions in the operation of both underground equipment and the wells themselves. Scheduled preventive maintenance is planned in advance and is carried out in accordance with repair schedules [68].

Restoration repair of wells is a repair caused by an unforeseen sharp deterioration in the technological mode of operation of wells or their shutdown due to pump failure, breakage of the rod column, etc. [68].

The routine repair of wells (RRW) is a complex of works aimed at restoring the operability of borehole and wellhead equipment, and work on changing the operating mode of the well, as well as cleaning the borehole equipment, the walls of the well and the bottom of various deposits (paraffin, hydrate plugs, salts, corrosion products) [68].

Restoration repair of wells is carried out in case of complete failure of the RDP as a result of jamming of the plunger, destruction of valves, breakage of rods. Restorative repair is used when the methods of routine repairs do not lead to the desired result. RRW methods provide for repairs without extracting RDP from the well. RRW is used in case of failure of the plug-in pump from the lock support. To restore operability, it is enough to lower the polished rod some distance down until the pump is reattached. Also, during the jamming of valves with paraffin deposits, a technique is used to increase the number of double strokes or washing the well with a technological solution. The method of increasing the double strokes of the pump is to increase the flow rate of the liquid passing through the valve seat and the breakdown of deposits from its surface. This method is applicable for small deposits. Monitoring of changes in operation is carried out using a dynamograph. The dynamograph draws a dynamogram of the work, after analyzing which a conclusion is made about the effectiveness of the methods [68].

Specialized methods for repairing RDP and wells equipped with RDP are described in the following patents.

The method of repairing a downhole rod deep pump (RF Patent, RU 2282750 C1, IPC F04B 47/00 [69]) – the invention relates to pumping equipment, in particular to rod deep pumps used in the operation of oil wells, and can be used in their repair, which consists in restoring the gap between the cylinder and the plunger. To do this, determine the areas on the surface of the cylinder on which it is necessary to restore the wall thickness, adjustable dielectric gaskets are installed inside the cylinder, separating the restored area from the rest of the cylinder surface, then restore the wall thickness of each section with metal to the required values, for example, by electrochemical deposition of chromium. This method of pump repair allows not only to restore failed pumps, but also to improve the characteristics of existing ones by reducing the gap between the cylinder and the plunger to match its values with higher landing tolerance groups.

A method for repairing the production column of a producing well (RF Patent, RU 2730158 C1, IPC E21B 33/10 [70]) – the invention relates to a method for repairing the production column of a producing well. The technical result is an increase in the reliability of the repair of the production column of the producing well. The method of repairing the production column of a producing well includes identifying the interval of violation of the production column, lowering into the production column a column of elevator pipes with a lock support of a plug-in rod depth pump, equipped from bottom to top with a plug, a perforated pipe and a packer, planting a packer in the production column below the interval of its violation, lowering into the elevator column of pipes of the RDP plunger on the column the rods, the landing of the RDP plunger in the lock support and the start of the RDP in operation. After the detection of the violation interval, the specific pick-up rate of the violation of the operational column is determined. If the specific pick-up of the violation is g<0.5 m^3/(hMPa), then pre-drainage of the violation with acid in a volume of 1.5 m^3 is performed until the specific pick-up g>0.5 m^3/ (hMPa) is reached, then tamponing of the violation in the production column is performed. After tamponing, the inflow of fluid from the violation is determined by a decrease in the level in the production column.

If there is no inflow of liquid from the violation of the production column, then it is equipped with pumping equipment and the producing well is put into operation. If the inflow of liquid from the violation of the production column is no more than 10% of the flow rate and this allows the production well to be operated cost-effectively, then a bottom-up layout is assembled at the mouth of the production well: a plug, a perforated pipe, a packer, a column of elevator pipes with a lock support of the plug-in RDP, the layout is lowered into the production column. If the packer is planted hermetically, then the plug-in RDP is lowered into the column of elevator pipes and put it in the lock support of the column of elevator pipes. If the inflow of liquid from the violation of the production column is more than 10% and exceeds the value of the cost-effective operation of the producing well, then the plugging is repeated until the inflow of liquid from the violation is reduced by no more than 10% and the cost-effective operation of the producing well is achieved [70].

Conclusion

The design features of the pumping units under study provide ample opportunities for modification of the initial elements (cylinder, plunger, valve pair, filter). The established factors affecting the duration of the operability of installations ensure consistency in the development of an algorithm for improving the design. From the study of the causes of failures, the main causes of failures have been identified, depending on the concentration of mechanical impurities in the pumped oil-containing liquid. Such complex heavy inclusions as ARPD, as well as high-hard abrasives (quartz sand) have the greatest impact on the service life of working parts, friction pairs and sealed connections of installations.

Analysis of common pumping units such as centrifugal pump and rod deep pump showed that rod deep pumps are common in the wells of the Republic of Kazakhstan and are structurally outdated with low production efficiency. However, the peculiarity of operation in low-flow wells with aggressive media and high content of mechanical impurities, rod depth pumps have found wide application and require ensuring high efficiency of RDP.

Based on the above, it is most cost-effective to increase the mechanical and physical properties of the working surfaces of the parts of the «plunger-cylinder» pair involved in mutual friction, as well as to increase the duration of the effective tightness of the «seat-valve» pair connection.

Modern methods of hardening are based on spraying high-strength hard or ceramic materials on vulnerable surfaces of parts that have undergone maximum wear during operation. Also, an increase in the efficiency of filtering devices helps to reduce the surface wear of working surfaces.

Chapter 2. Gas dynamic analysis of a downhole rod deep pump

2.1 Overview of valve pair throughput

To analyze and visualize the processes occurring in the working space of the PP2B-44 rod depth pump (Figure 27), a hydrodynamic calculation was performed using Simulation Flow software based on the CAD editor of SolidWorks.

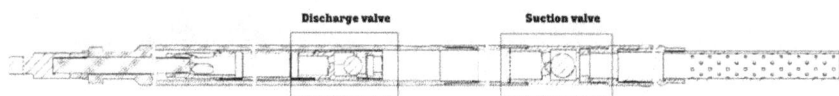

Figure 27. RDP valve pairs

The primary calculation consists of checking the geometry of the pump discharge valve parts for the created resistance to the flow of liquid during the movement of the plunger part downwards. Figure 28 shows the valve assembly of the plunger assembly.

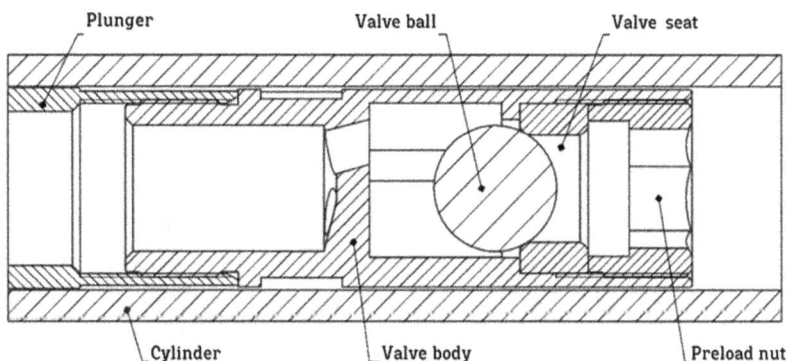

Figure 28. Valve assembly of the plunger assembly (discharge valve)

The "Project Wizard" indicates the input parameters for calculating oil flows, in the area of the plunger valve, with an internal flow. To begin the calculation, the boundary values of the pressure at the plunger outlet were introduced, which is equal to the pressure of the liquid column at a depth of 2,000 m – 20 MPa and the average volume flow through the valve seat (Figure 29) of the PP2B-44 pump is 17 m³/day or in the second equivalent of the volume flow of 0.0002 m³/sec.

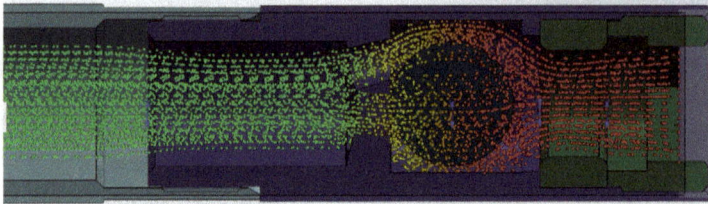

Figure 29. Distribution of compressed fluid flow along the trajectory

From the visualization of the preliminary calculation of the fluid flow (Figure 29), it can be seen that the main resistance is caused by the compression of the fluid flow when passing through three holes and the gap formed between the ball of the discharge valve Ø26.988 with the valve body. The formed cross-sectional area in the first case (Figure 30, Figure 31) is 235.5 mm², and in the second 286.2 mm². From the comparison it can be seen that the holes (Figure 30) have greater resistance when the flow passes through them. To clarify the effect of the holes on the flow, it is necessary to carry out an additional calculation of the velocity of fluid movement through the valve (Figure 32).

Figure 30. Total cross-sectional area of the holes in the valve body

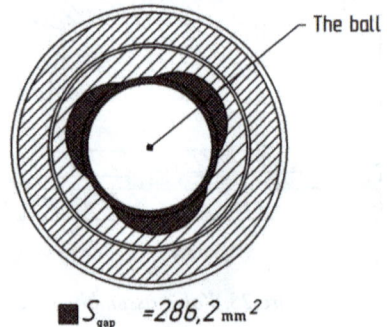

Figure 31. Cross-sectional area of the gap between the ball and the valve body

An additional calculation will allow you to narrow down the zones for searching for critical places with possible cavitation. The main condition for the formation of cavitation is a sharp change in the velocity and pressure of the liquid, which lead to the appearance of vapor bubbles in the liquid having a destructive effect on the surface of the parts when collapsing. The study will determine the presence of cavitation destruction on the walls of the working cylinder of the RDP.

Visualization of the change in flow velocity shows where the main compression of the flow occurs (Figure 32).

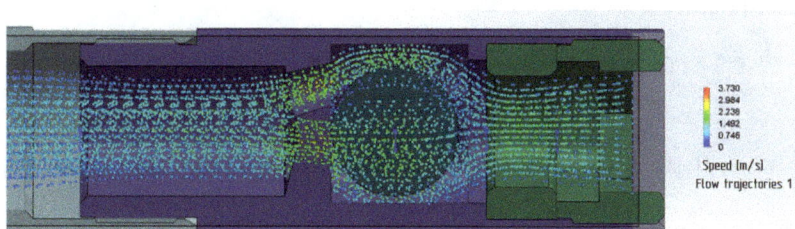

Figure 32. Visualization of the flow velocity of an oily liquid

From the representation of the trajectory, it can be seen that in the pockets between the body and the ball, the flow velocity increases to 2.238 m/s, and in the holes the flow moves in the velocity interval from 2.238 m/s to 2.984 m/s. The speed is limited by the capacity of the gaps and the shape of the holes.

To clarify the results obtained, it is necessary to carry out an additional hydrodynamic calculation.

2.2 Preliminary calculation of RDP operating parameters

In order to determine the exact hydrodynamic characteristics, it is necessary to carry out an updated calculation of the operational parameters of the pump. A rod plug–in pump of mechanical fastening with a lower lock arrangement – 73-PP1B-44-12-18 was adopted as the basis for the beginning of the study (Table 4). Determination of the parameters of the flow of oil-containing liquid in the working space of the pumping unit is necessary to carry out a clarifying program calculation in the FloEFD software module.

Table 4. Characteristics of the PP1B-44 pump [31]

Pump characteristics	Pump code	Conditional size	Plunger working stroke, [mm]	Plunger length, [mm]	Pressure, [m]	Rod thread
Plug-in with mechanical fastening and lower lock location	73-PP1B-44-12-18	44	1,500	1,200	1,600	Ш-19

The supply of the deep - pump unit Q_{liquid} of the rod pump is determined by the following formula [71]:

$$Q_{liquid} = 1,440 \cdot \frac{\pi \cdot D^2}{4} \cdot S_p \cdot n \cdot \rho_l \cdot \eta, \qquad m^3/day, \qquad (1)$$

where D – the diameter of the pump plunger (conditional size), m;

S_p – working stroke of the plunger, m;

n – number of swings (double strokes) per minute;

ρ_l – relative density of liquid (oil), ρ_l=0.85 g/cm³;

η – the feed coefficient of the pumping unit.

Overall, the feed coefficient:

$$\eta = \eta_1 + \eta_2 \qquad (2)$$

Feed coefficient that takes into account the elongation of the rods during the double stroke [71]:

$$\eta_1 = 1 - \frac{10^5 \cdot P_l \cdot L}{S_p \cdot E}\left(\frac{1}{f_r} + \frac{1}{f_p}\right), \qquad (3)$$

where P_l – weight of the liquid column above the pump plunger, kg;

L – the pump descent depth, m;

S_p – plunger stroke length, m;

E – the modulus of elasticity of the metal, Pa;

f_r – the cross-sectional area of pumping rods, cm²;

f_p – the cross-sectional area of pumping pipes, cm².

$$\eta_1 = 1 - \frac{10^5 \cdot 2{,}025 \cdot 1{,}600}{1.5 \cdot 200{,}000 \cdot 10^6} \cdot (0.38) = 0.59.$$

The weight of the liquid column above the pump plunger, N [71]:

$$P_l = \frac{L \cdot \rho_l \cdot g \cdot F}{10^4},\tag{4}$$

where L – the pump descent depth, m;

ρ_l – liquid density, g/cm^3;

g – acceleration of free fall, g=9.8 m/s^2;

F – the cross–sectional area of the pump plunger, cm^2.

$$P_{liquid} = \frac{160{,}000 \cdot 0.85 \cdot 9.8 \cdot 15.2}{10^4} = 2{,}025 \text{ kg.}$$

The coefficient η_2 is determined by the nomogram in Figure 33. In the nomogram, the pump installation depth L=1,600 m and n=15 dv. stroke/min are used to determine η_2 according to the characteristics of the rocking machine in Table 5.

Table 5. Characteristics of the rocking machine [72]

The cipher of the rocking machine	Indicators		
	Rated load (on the wellhead rod), [kN]	Nominal stroke length of the wellhead rod, [m]	The number of moves of the balancer per minute
RMD3-1.5-710	30	1.5	5 – 15

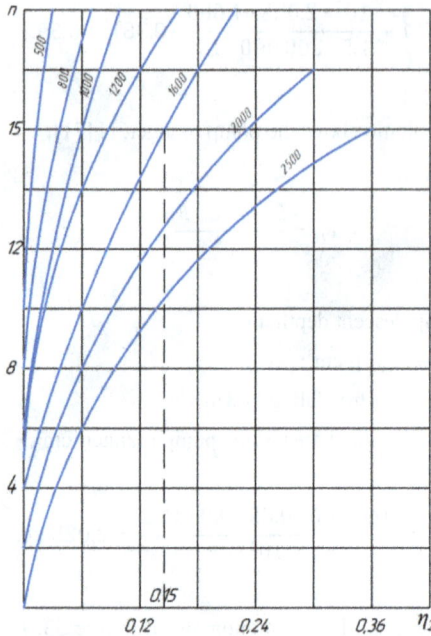

Figure 33. Nomogram of the dependence of the coefficient η_2 on the number of double strokes of the rocking machine [71]

According to the nomogram, the value of the coefficient η_2=0.15. In this case, the value of the pump feed coefficient η:

$$\eta = 0.59 + 0.15 = 0.74.$$

Based on the above calculation of the pump feed coefficient η, it is possible to calculate the feed of the rod depth pump Q_{liquid}:

$$Q_{liquid} = 1,440 \cdot \frac{3.14 \cdot 0.044^2}{4} \cdot 1.5 \cdot 15 \cdot 0.85 \cdot 0.74 = 30 \, \text{m}^3/\text{day}.$$

The refined calculation showed that when installing a rod depth pump to a depth of 1,600 m, the liquid flow rate was 30 m³/day. The increase in Q_{liquid} production, relative to the

Materials Research Forum LLC
https://doi.org/10.21741/9781644902356

installation of the pump to a depth of 2,000 m, is due to the fact that with a decrease in the total length of the rods, the rigidity of the system increases and it becomes possible to install more (*n*) double plunger strokes per minute. 15 strokes of the rocking balancer per minute corresponds to a plunger movement speed of 45 m/min, which corresponds to a movement speed of 0.75 m/sec.

For further program calculation, it is necessary to determine the pressure at the pump outlet p_{po}. As the pressure at the pump outlet, we take the pressure of the liquid column at a depth of 1,600 m. In this case, the internal pressure at the pump outlet will be equal to p_{po} = 13.34 MPa.

The values given above correspond to the parameters of an ideal fluid. In this regard, an additional adjustment calculation of indicators was made, taking into account the characteristics of the pumped products.

For the correction calculations, the averaged physico-chemical characteristics of high-paraffin oil with a high gas content lying on the territory of the Republic of Kazakhstan were taken (Table 6) [73]. This type of oil liquid is characterized by the complexity of production due to increased wear of equipment from contained mechanical impurities.

Table 6. Physico-mechanical characteristics of oil [73]

Physico-chemical indicators	Values
Density, [kg/m³]	850
Viscosity at 20 °C, [m²/s]	65.18·10⁻⁴
Gas content in oil, [m³/m³]	111.89
Content, wt. [%]:	
Paraffin	17.56
Resin	9.56
Asphaltene	2.88
Gas content in oil, [m³/m³]	111.89
Thermobaric conditions:	
Reservoir temperature, [°C]	82.09
Reservoir pressure, [MPa]	22.75

The calculation of the physical properties of oil under given conditions is carried out according to the following formulas [74]:

The amount of gas dissolved in oil $G_0(p)$ is determined by the formula:

$$G_0(p) = G_0(p_{sat})[(p - p_0)/(p_{sat} - p_0)]^c, \quad (5)$$

where $G_0(p_{sat})$ – the amount of gas dissolved in 1 m^3 of oil at saturation pressure (p_{sat}), reduced to normal conditions, m^3/m^3;

p, p_0 – respectively the current and atmospheric pressure, $p_{sat} \le p \le p_0$, MPa;

c – an empirical coefficient, the value of which for further calculations is assumed to be equal to $c = 0.5$.

The volume coefficients of oil $b_{oil}(p)$ and liquid $b_l(p)$ are calculated using the following formulas [74]:

$$b_{oil}(p) = 1 + (b_{oil} - 1)[(p - p_0)/(p_{sat} - p_0)]^{0.25} \qquad (6)$$

$$b_l(p) = b_{oil}(p)(1 - \beta_{water}) + b_{water}(p)\beta_{water} \qquad (7)$$

where b_{oil}, $b_{water}(p)$ – the volume coefficient of oil at $p = p_{sat}$ and water, respectively. In further calculations, it is assumed that $b_{water}(p) = 1$, b_{oil} is determined by the ratio of the density of oil at atmospheric pressure and the density of reservoir oil at reservoir temperature according to the nomogram (Figure 34) [75].

The volume coefficient is equal to:

$$b_{oil} = 850/830 = 1.024.$$

The diagram shows that with an increase in the temperature of the liquid in the oil-bearing formations, the density decreases. Figure 34 shows that the density of 869 kg/m^3 is closest to the density 850 kg/m^3 used for calculation and the density value of the heated oil liquid is selected according to the approximate curve in the diagram.

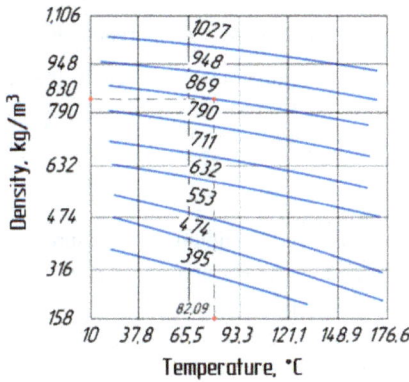

Figure 34. Diagram of density change from temperature

To calculate the flow characteristics of a mixture of liquid and gas, at the current pressure p in the pump sections, it is necessary to determine the following parameters [74]:

    – liquid flow rate, m³/s:

$$Q_{liquid}(p) = Q_{oila}b_l(p)(1 - \beta_{water}), \qquad (8)$$

    – free gas consumption, m³/m³:

$$V_{con.\ gas}(p) = [G_0(p_{sat}) - G_0(p)]\, zp_0T_{well}Q_{oila}/(pT_0), \qquad (9)$$

    – consumption of gas-liquid mixture, m³/m³:

$$Q_{mix}(p) = Q_{liquid}(p) + V_{con\ gas}(p), \qquad (10)$$

    – density of gas-saturated oil, kg/m³:

$$\rho_{oil}(p) = [\rho_{oila} + \rho_{gas}\, G_0(p)]/b_{oil}(p), \qquad (11)$$

where $Q_{oila} = Q_l(1 - \beta_{water})$ – flow rate of degassed oil, m³/s;

49

$\beta_{water} = 0.1$; $T_0 = 273$ K;

T_{well} – the average temperature in the borehole, K;

z – the gas super-compressibility coefficient, the value of which in further calculations is assumed to be $z=1$;

ρ_{gas} – gas density, kg/m^3.

To determine the flow of a gas-saturated liquid, the pressure value at the pump intake is necessary. To calculate the pressure at the pump intake (p_{pi}), it is recommended to take equal to 30% of the gas saturation pressure (p_{sat}). In this case, the gas saturation pressure is calculated based on the free gas content in oil (m^3/m^3) and the coefficient A. The coefficient A depends on the density of the produced products and for $\rho = 0.85$ g/cm^3 is equal to $A = 0.1293$.

Gas saturation pressure:

$$p_{sat} = AG, \tag{12}$$

where G – the gas content in oil, m^3/m^3.

$$p_{sat} = 0.1293 \cdot 111.89 = 14.46 \text{ MPa.}$$

Pressure at the pump intake:

$$p_{pi} = 0.3 \cdot 14.46 = 4.34 \text{ MPa.}$$

We determine the flow rate of the gas-liquid mixture at a pressure of p_{pi}:

$$Q_{oild} = 3.5 \cdot 10^{-4} \cdot (1 - 0.1) = 2.7 \cdot 10^{-4} \text{ m}^3\text{/s;}$$

$$b_{oil}(p_{pi}) = 1 + [(4.34 - 0.1)/(14.46 - 0.1)]^{0.25} = 1.22;$$

$$l(p_{pi}) = 1.22 \cdot (1 - 0.1) + 1 \cdot 0.1 = 1.198;$$

$$Q_{liquid}(p_{pi}) = 2.7 \cdot 10^{-4} \cdot 1.198 \cdot (1 - 0.1) = 2.9 \cdot 10^{-4} \text{ m}^3/\text{s};$$

$$G_0(p_{pi}) = 111.89 \cdot [(4.34 - 0.1) / (14.46 - 0.1)]^{0.5} = 60.8 \text{ m}^3/\text{m}^3;$$
$$V_{con.\ gas}(p_{pi}) = [111.89 - 60.8] \cdot 0.1 \cdot 324 \cdot 2.7 \cdot 10^{-4}/(4.34 \cdot 273) = 3.8 \cdot$$
$$\cdot 10^{-4} \text{ m}^3/\text{m}^3;$$

$$Q_{mix}(p_{pi}) = 2.9 \cdot 10^{-4} + 3.8 \cdot 10^{-4} = 6.7 \cdot 10^{-4} \text{ m}^3/\text{s} = 58 \text{ m}^3/\text{day};$$

2.3 Calculation of RDP performance indicators in the production of multicomponent oil

The gas separation coefficient at the reception of the RDP is determined by the approximate formula [74]:

$$\sigma_s = \frac{1 - (D_{PCP}/D_{pc})}{1 + 0.93 \cdot Q_{liquid}(p_{pi})/\omega_s \cdot D_{pc}^2}, \tag{13}$$

where D_{PCP} – the outer diameter of pumping-compressor pipes (PCP) at the pump
intake, mm;

D_{pc} – the inner diameter of the production column well, m;

$Q_{liquid}(p_{pi})$ – liquid flow rate, m³/s;

ω_s – the relative velocity of gas movement at the pump reception area, at
$\beta<0.5$ we assume equal to 0.02 m/s.

$$\sigma_s = \frac{1 - (0.06/0.15)}{1 + 0.93 \cdot 2.9 \cdot 10^{-4}/0.02 \cdot 0.15^2} = 0.375.$$

Due to the separation of a part of the free gas at the pump intake, according to the gas separation coefficient, the gas factor of the liquid changes, which is calculated by the formula [74]:

$$G_0' = G_0(p_{sat}) - [G_0(p_{sat}) - G_0(p_{pi})]\sigma_s, \tag{14}$$
$$G_0' = 111.89 - [111.89 - 60.8] \cdot 0.375 = 92.73 \text{ m}^3/\text{m}^3.$$

The correction value of the saturation pressure p_{sat} in accordance with the pipe gas factor is calculated by the formula [74]:

$$p'_{sat} = (G'_0/G_0(p_{sat}))^{1/c} \cdot (p_{sat} - p_0) + p_0. \tag{15}$$

$$p'_{sat} = (92.73/111.89)^{1/0.5} \cdot (14.46 - 0.1) + 0.1 = 9.99 \text{ MPa}.$$

Calculation of the amount of free gas entering the pump $V'_{con.gas}$ and the gas-liquid mixture $Q'_{mix}(p_{pi})$, at saturation pressure, is calculated according to the formulas [74]:

$$V'_{con.gas}(p_{pi}) = V_{con.gas}(p_{pi}) \cdot (1 - \sigma_s), \tag{16}$$

$$V'_{con.gas}(p_{pi}) = 3.8 \cdot 10^{-4} \cdot (1 - 0.375) = 2.4 \cdot 10^{-4} \text{ m}^3/\text{m}^3.$$

$$Q'_{mix}(p_{pi}) = Q_{liquid}(p_{pi}) + V'_{con.gas}(p_{pi}), \tag{17}$$

$$Q'_{mix}(p_{pi}) = 2.9 \cdot 10^{-4} + 2.4 \cdot 10^{-4} = 5.3 \cdot 10^{-4} \text{ m}^3/\text{s} = 46 \text{ m}^3/\text{day}.$$

Next, we calculate the oil column at the pressure p_{po} at the pump outlet using the standard formulas given above.

$$b_{oil}(p_{po}) = 1 + (1.3 - 1)[(13.34 - 0.1)/(14.46 - 0.1)]^{0.25} = 1.3;$$

$$b_l(p_{po}) = 1.3 \cdot (1 - 0.1) + 1 \cdot 0.1 = 1.3;$$

$$Q_{liquid}(p_{po}) = 2.7 \cdot 10^{-4} \cdot 1.3 \cdot (1 - 0.1) = 3.2 \cdot 10^{-4} \text{ m}^3/\text{s};$$

$$G_0(p_{po}) = 111.89 \cdot [(13.34 - 0.1)/(14.46 - 0.1)]^{0.5} = 107.4 \text{ m}^3/\text{m}^3;$$

$$V'_{con.\ gas}(p_{po}) = [111.89 - 107.4] \cdot 0.1 \cdot 324 \cdot 2.7 \cdot 10^{-4}/(13.34 \cdot 273) = 0.1 \cdot \\ \cdot 10^{-4}\ \mathrm{m^3/s};$$

$$Q'_{mix}(p_{po}) = 3.2 \cdot 10^{-4} + 0.1 \cdot 10^{-4} = 3.3 \cdot 10^{-4}\ \mathrm{m^3/s} = 28.5\ \mathrm{m^3/day}.$$

Using the output values, it is possible to calculate the operating pressures ($p_{suc.\ c}$, p_{dc}) in the cylinder and the flow rate ($Q_{suc.\ v}$, Q_{dv}) through the valves during the operation of the suction and discharge valves.

Fluid flow rates through the valves correspond to the following formulas [74]:

$$Q_{suc.v} = Q'_{mix}(p_{pi})\ \text{и}\ Q_{dv} = Q'_{mix}(p_{po}). \tag{18}$$

The maximum speed of the product movement (v_{max}) through the valve seat opening and the Reynolds number (Re_v) corresponding to this speed are calculated according to the formulas [74]:

$$v_{max} = \frac{\pi Q_v}{S_v}, \tag{19}$$

$$Re_v = \frac{v_{max}d_v}{v_l}, \tag{20}$$

where S_v – the cross-sectional area of the seat opening, m;

v_l – the kinematic viscosity of the liquid, $\mathrm{m^2/s}$.

The viscosity of the largest component from the pumped liquid (water, oil, etc.) is selected as v_l. The average viscosity in Kazakhstan deposits at a temperature $T = 330$ K varies from $0.03 \cdot 10^{-4}\ \mathrm{m^2/s}$ to $0.34 \cdot 10^{-4}\ \mathrm{m^2/s}$ for calculation, we take the value $v_l = 0.2$.

$$v_{max\ suc} = \frac{3.14 \cdot 5.3 \cdot 10^{-4}}{3.14 \cdot 0.030^2} = 0.6\ \mathrm{m/s},$$

$$v_{max\,d} = \frac{3.14 \cdot 3.3 \cdot 10^{-4}}{3.14 \cdot 0.023^2} = 0.63 \text{ m/s},$$

$$Re_{suc.v} = \frac{0.6 \cdot 0.03}{0.2 \cdot 10^{-4}} = 900,$$

$$Re_{d\,v} = \frac{0.63 \cdot 0.023}{0.2 \cdot 10^{-4}} = 724.5.$$

The pressure difference before and after the valve is calculated by the formula [74]:

$$\Delta p_v = \frac{v_{max}^2 \cdot \rho_{ld}}{2\xi_v^2}, \text{MPa} \tag{21}$$

where ξ_v – the coefficient of resistance to the flow of liquid, depending on the pressure, flow rate and degree of valve opening. For the calculation, we take $\xi_v = 0.4$;

ρ_{ld} – the density of the degassed liquid, kg/m³.

In this case, the pressure drop Δp_v in the pump valves is calculated according to the formula [74]:

$$\rho_{ld} = \rho_{oild}(1 - \beta_{water}) + \rho_{water}\beta_{water}, \tag{22}$$

where ρ_{ld} – density of degassed oil, kg/m³.

$$\rho_{ld} = 850(1 - 0.1) + 1,000 \cdot 0.1 = 860 \text{ kg/m}^3;$$

$$\Delta p_{suc.\,v} = \frac{0.6^2 \cdot 860}{2 \cdot 0.4_{suc.\,v}^2} = 0.00097 \text{ MPa},$$

$$\Delta p_{d\,v} = \frac{0.63^2 \cdot 860}{2 \cdot 0.4_{d\,v}^2} = 0.001 \text{ MPa}.$$

The pressure drop determines how the pressure in the sealed area will change after passing through the valve. The pressure in the zones, taking into account the differential during suction and discharge, is determined by the following formulas:

$$p_{suc.\ c} = p_{pi} - \Delta p_{suc.\ v} = 4.34 - 0.001 = 4.399 \text{ MPa};$$

$$p_{d\ c} = p_{po} - \Delta p_{d\ v} = 13.34 - 0.001 = 13.339 \text{ MPa}.$$

For hydrodynamic calculation of the flow of oil-containing liquid in the working chamber, software modeling in the FloEFD environment of the dynamics of the process of passing oil and gas liquid through the suction valve of the RDP was used (Figure 35). For this calculation, the actual volumetric flow rate $Q_{suc.\ v}$ at the valve inlet is equal to $Q_{suc.\ v} = 5.3 \cdot 10^{-4}$ m^3/s and static pressure $p_{suc.\ c} = 4.399$ MPa in the flow section of the valve seat opening.

Applying the calculated values for hydrodynamic calculation, the following values of pressure (p) and fluid velocity (v) are obtained, indicated by the isosurface in Figure 35 and Figure 36.

Graphical representation and calculation in the FloEFD software environment showed that a mechanically complex oil and gas liquid, when passing through the valve, acquires a velocity of $v_{suc} = 0.75$ m/s and is consistent with the calculated velocity value of $v_{suc} = 0.6$ m/s. From the presented model (Figure 35), it was found that the static pressure ($p_{suc.\ c}$) does not have a significant effect on the velocity of fluid movement $v_{suc} = 0.75$ m/s through the valve seat opening. Modeling has proved that the main factors shaping the high-speed flow are the change in the values of the liquid flow $Q_{suc.\ v}$ and the diameter of the hole in the valve seat equal to $d_{suc} = 30$ mm. From the graphical representation, it can be seen that the increase in the speed mode $v_{suc.\ v} = 0.78$ m/s occurs in the gap between the pocket of the valve body and the valve ball and increases to $v_{suc.\ v} = 1.27$ m/s closer to the flow section of the valve. Further, in the valve space, the alignment and stabilization of the fluid velocity regime is observed to the values of $v_{suc.\ v} = 0.75 - 0.78$ m/s. The transition zones of the cross-section of the valve outlet openings contribute to the expansion of the flow and the change in pressure $p_{suc.\ c} = 4,426,276$ Pa.

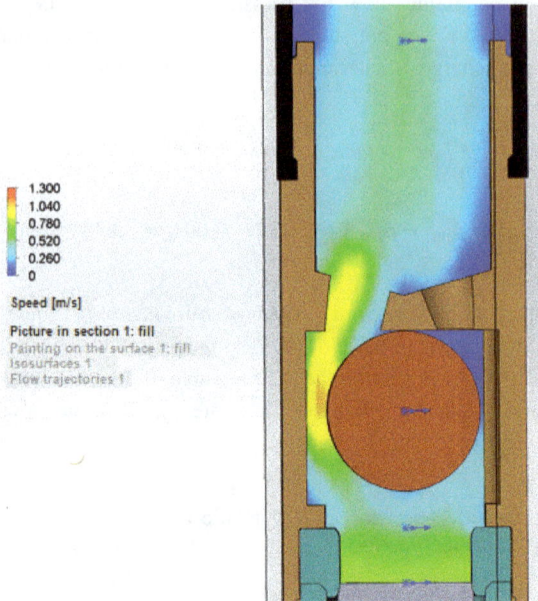

Figure 35. Fluid velocity distribution without cavitation calculation

The described process focuses the convergence area of three fluid flows at the outlet of the valve and creates turbulent flows that carry multicomponent mechanical and abrasive impurities along the inner surfaces of the extension. The constantly increasing layer of ARPD on the inner surfaces of the extension cord and PCP reduces the efficiency of extraction. During the operation of the RDP, the thickness of the ARPD layer increases, which reduces the technological gaps between the working surface of the plunger and the extension cord. The reciprocating motion and suction effect of the plunger constantly increases the ARPD layer and then inevitably compacts it. This effect contributes to the breakdown of the stuck layer and its transfer to the friction zone of the plunger-cylinder pair, which leads to accelerated abrasive wear and coking of the extension cord and PCP. This explains the mechanism and the area of wear of RDP parts.

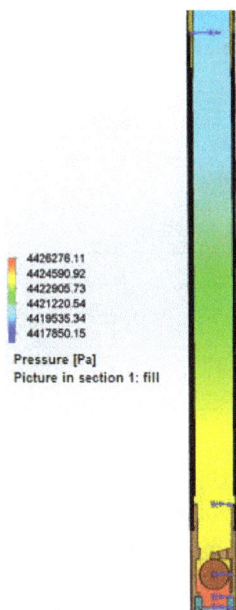

Figure 36. Pressure distribution in the suction valve area without cavitation

Figure 36 shows the pressure change from the valve seat to the pump plunger. On the isosurface it can be seen that the pressure increases after passing through the hole $d_{suc.\ v}$= 30 mm, but subsequently, when passing the gaps between the ball and the valve body, the flow narrows while increasing its speed and the pressure on the valve walls decreases. Further approaching the plunger, the pressure decreases. The decrease in pressure is associated with the movement of the plunger and, as a consequence, the creation of a suction effect. This calculation made it possible to determine small deviations of the pressure created by the flow of the inner walls of the valve and cylinder $p_{suc.\ c.\ program}$ from 4,417,850 Pa to 4,426,276 Pa relative to the design pressure $p_{suc.\ c}$= 4.399 MPa.

The calculations given above were carried out without an activated liquid cavitation accounting module, the parameters of the medium were set to calculate cavitation (Table 7).

Table 7. Oil and gas fluid parameters

Property	Meaning
Name	oil
Density	850 kg/m^3
Dynamic viscosity	0.128 Pa·s
Specific heat (C_p)	2,100 J/(kg·K)
Coefficient of thermal conductivity	0.12 W/(m·K)
Cavitation effect	+
Molar mass	0.3 kg/mol
Temperature	330 K
Saturation pressure	14,400,000 Pa
Radiation properties	−

The calculation of the fluid velocity under the conditions of cavitation effects showed (Figure 37) that the fluid located after the seat of the ball forms a backflow region and moves to the valve seat. Also, in the area of the seat opening, there is an area with sharp changes in the flow rate up to 80 m/s, as well as a reverse flow of liquid, which is impossible under the conditions of the suction stroke. The location of the increased velocity, as well as the areas with the reverse flow flow, are caused by the inaccuracy of processing the input data by the program due to the absence of cavitation in this area of the real fluid flow at the specified parameters. Additionally, the liquid pressure was calculated (Figure 38) with the included cavitation calculation module in the FloEFD program. The calculation results are presented by an isosurface and an arrow distribution of flows to visualize the outflow of liquid inside the valve assembly due to the most likely place of cavitation formation.

It can be seen from Figure 38 that the fluid flows move in the opposite direction from the actual direction of the plunger supply, which is impossible under the conditions of the suction stroke. This phenomenon indicates an error in the software calculation due to the absence of cavitation in the study area. Also, the phenomenon is caused by the high gas content of dissolved gas in the liquid and when the liquid enters the working area of the pump, the saturation pressure decreases. The reduced saturation pressure causes natural gas release, which is not cavitation and does not have a destructive effect on the parts of the rod pump.

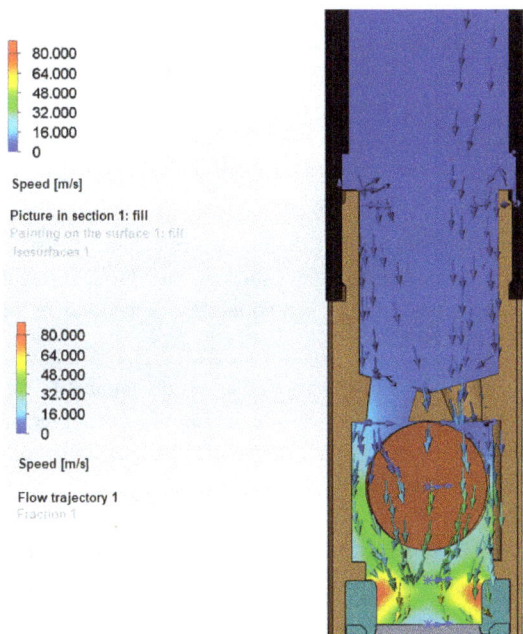

Figure 37. Distribution of fluid flow velocities taking into account cavitation

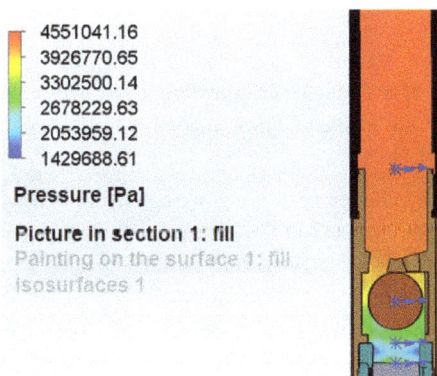

Figure 38. Pressure distribution taking into account the cavitation effect

The phenomenon of cavitation can occur only locally in places with rapid changes in velocity or pressure indicators. On the isosurface (Figure 38), there is a reduced fluid pressure in the area of the seat opening to 1,429,688 Pa. A decrease in liquid pressure when cavitation occurs in this area is possible, since a large accumulation of gas balls is formed that create a discharged mixture, but the most likely area of cavitation in the valve is the area of narrowing of the flow in the area of the holes and on the walls in the places of the greatest increase in the fluid flow rate.

There is no sharp pressure and velocity drop on the calculated isosurface (Figure 38), and therefore it can be concluded that cavitation in the FloEFD medium is calculated incorrectly, since it is impossible for cavitation regions to occur in the fluid flow at the specified parameters (*Table 7*).

The factor of uneven filling of the pump, defined as the amount of free gas in the pump $V'_{con.\ gas}(p_{pi}) = 2.4\cdot10^{-4}$, creates an area of damping of sudden pressure drops inside the working chamber of the pump, which eliminates the possibility of cavitation, which reduces the proportion of cavitation wear.

2.4 Research of modern technologies of recovery of mining pumps

Investigation of the causes of failure of the RDP revealed that during operation, the surface of the cylinder of the rod pump is more susceptible to wear compared to other parts of this system of contacting parts (a pair of cylinder-plunger). This is due to the fact that a coating consisting of a powder coating with a hard alloy and having a thickness of up to 2 mm is applied to the surface of the plunger, and the surface of the cylinder is subjected to chemical-thermal treatment with a thickness of the hardened layer up to 0.5 mm. In this regard, a thin nitrided layer on the inner surface of the cylinder may be damaged, in the form of a shell, which from the inside will not be protected from the corrosive properties of the medium. Thus, a small damage leads to accelerated wear of the cylinder surface.

Based on the above, the main reason leading to irreversible wear of the movable plunger-cylinder pair is corrosion-mechanical wear, as a result of contact of working, precision-treated surfaces, with mechanical impurities, which include paraffins (17.6%), resins (9.56%), asphaltenes (2.88%), as well as ingress into the gap between the surfaces of friction, solid inclusions and accompanying sand acting as an aggressive abrasive of high hardness. According to the Mohs scale, the hardness of quartz sand corresponds to 7 degrees of hardness and is similar to 60 – 70 HRC on the Rockwell scale [76].

According to GOST 31835-2012 [31], the method of chemical-thermal nitriding treatment of the RDP cylinder allows to achieve a hardness of 65 – 66 HRC for the cylinder material (Steel 38Cr2MoAlA) at a depth of the nitrided layer up to 0.5 mm, which does not allow to fully exclude the possibility of damage to the hardened layer, and also completely excludes the possibility of restoring the worn surface by this method.

A wide variety of factors that catastrophically reduce the durability of deep rod pumps necessitates the development of an optimal universal method for restoring critical internal surfaces of small diameter and long length. Thus, there is a complex scientific and technical problem of developing a unique method for restoring the inner surface of a small-diameter rod pump that provides high physical and mechanical properties of the coating and manufacturability. The problem of restoration is focused on the elimination of mechanical wear in the form of a change in the design geometry of the surface and the restoration of the phase structure of a carbide material with corrosion-resistant properties.

To solve this problem, it is necessary to solve the main tasks:
- to investigate the technological features of existing methods of RDP recovery;
- to develop a design to restore the physical-mechanical properties of metal pump products;
- to develop a technology for restoring the internal surfaces of small diameter RDP;
- to justify the optimal modes of laser recovery;
- to substantiate optimal materials that provide high physical-mechanical properties of the modified surface.

The main requirements for RDP of low-flow wells are their compliance with the requirements of drawings, high physical and mechanical properties (strength, fatigue resistance, fretting resistance, erosion and corrosion resistance), specified productivity [33, 34, 77].

Researchers Yu.I. Blinov, V.A. Shurinov, S.L. Chernyshevich, V.V. Yakovlev, V.R. Fedorin developed a method for manufacturing a rod pump [78, 79] by applying a system of recesses, before the final operation of finishing, removal of damaged material, thermal and mechanical stress treatment.

The disadvantage of the method is the unacceptability of applying depressions by chemical methods, the technical complexity of processing the inner surface of long-length cylinders of high-strength nitrided steel 38Cr2MoAlA, and mechanical methods of depressions lead to the formation of burrs and surges, reducing the quality and operability of the rubbing surfaces of parts, increasing the run-in time.

The method of manufacturing a pumping rod (RF Patent, RU 2119858 C1, [80]), proposed by O.R.Valiakhmetov, G. Richmuller, R.R. Mulyukov (81, 82), includes the use of blanks made of steel of different grades for the body and heads of the rod, thermal and mechanical processing of the blanks is carried out separately, then by friction welding to the ends of the rod, the heads are welded.

The disadvantages are poor quality and stress concentration in the rod. Due to the high heterogeneity of the structural and phase composition at the welding site, the rods break off. The method is not inferior in labor intensity and energy intensity to the method [44], which requires special equipment that welds long parts Ø8 – 30 mm with a force of 10 tons.

Tensile stresses reduce the fatigue resistance of the material, and the structural-phase heterogeneity of the material reduces its corrosion resistance, since the combination of sites with different electrode potential forms many microgalvanopares.

Problems with the hardening of the plunger are currently being solved by spraying a hard alloy with high chemical and mechanical wear resistance on its surface. The experience of operating domestic RDP has shown that nitriding and chrome plating give the best results in wear resistance of cylinders.

Hardening coating – spraying of the outer surface with carbide powders. The base of the powder is nickel (70%), which causes high corrosion resistance. The chromium content in the powder (15 – 18%) causes a high coating hardness of 56 – 65 HRC. Therefore, these plungers are used in various environments, both with the content of abrasive particles (sand, scale, etc.), and corrosive environments with the presence of hydrogen sulfide (H_2S), and carbon dioxide (CO_2). The thickness of the coating must be at least 0.35 mm to maintain hardness during wear during work. At the same time, a significant increase in the thickness of the coating (more than 0.6 mm) will lead to its low adhesion to the substrate and chips during operation.

The advantage of chrome plating is in the high hardness and wear resistance of the coating, as well as in the absence of warping (bending) of the cylinder when applying this coating. The disadvantage is the porosity of any chrome coating and the small thickness of the layer. Corrosion tests of the chrome coating show that according to the existing defects, intensive etching occurs, both in width and in depth (Figure 39). In addition, the chromium plating process is dangerous to health, 6-valent chromium causes cancer.

Figure 39. Foci of corrosion of the base metal of the inner surface of the cylinder as a result of the chrome plating process

The advantages of nitriding include the high hardness and wear resistance (including corrosion) of the nitride layer, the absence of porosity under certain conditions, and the relatively large thickness of the nitrided layer. However, when restoring the cylinder, warping takes place – its transverse bending. This is due to the difference in the thickness of the cylinder itself and the unevenness of the nitrided layer along the diameter of the channel (Figure 40).

The traditional technology of ion-vacuum nitriding (IVN) with a depth of $0.25 - 0.30$ mm increases the non-straightness of the axis of the cylinder section with a length of 1m to an average of 0.2 mm, which exceeds the tolerance for this parameter by 2 times. To ensure a tolerance of 0.1 mm regulated by API and GOST standards, a transverse bending correction is required, after which cracks of $2 - 5$ microns appear on the nitrided surface.

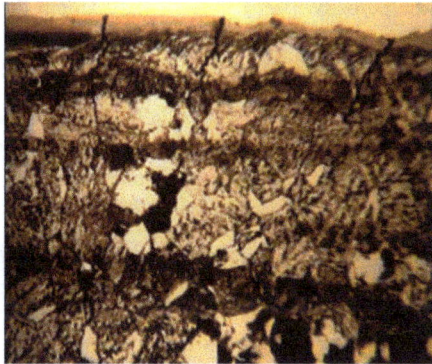

Figure 40. The appearance of cracks on the nitrided surface after bending correction

In addition, after editing, to ensure a guaranteed clearance between the cylinder and the plunger of 0.025 mm, honing with multi-row honing heads having a rigid body is necessary. Such honing ensures straightness of the channel axis, but removes part of the nitrided layer up to $0.03 - 0.1$ mm deep.

The advantages of chrome plating are high hardness (Figure 41) and wear resistance of the coating, as well as the absence of warping (bending) of the cylinder when applying this coating.

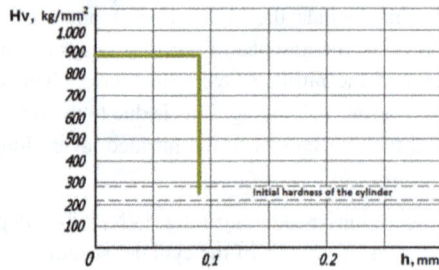

Figure 41. Hardness distribution over the depth of the chrome-plated cylinder layer

The disadvantage is the porosity of any chrome coating and the small thickness of the layer. Corrosion tests of the chrome coating show that according to the existing defects, intensive etching occurs, both in width and in depth of the metal (Figure 42).

After seven days, the dimensions of the defects in depth exceed the thickness of the coating, that is, corrosion of the base metal begins. Comparative corrosion tests of the nitrided layer with an allowance of 2 – 3 microns and 50 microns showed an increase in the corrosion rate by 3 times, from 0.12 g/m^2h to 0.37 g/m^2h.

Therefore, an important scientific-technical task is the development of an energy-efficient technology for manufacturing a pumping complex with high physical-mechanical properties of the parts of the RDP well.

Figure 42. The state of the chrome coating of steel samples after exposure in reservoir water for 7 days at T = 60 °C

Plasma technologies for the manufacture of oil-producing equipment using ionized gas, multicomposite powders and wires are widely used in Europe [83 – 91].

Strict requirements for the RDP of low-flow wells are regulated by corrosion-resistant, wear-resistant nickel and cobalt alloys ZMI-ZU, ZHS-6, U-5000, Fsx-414, etc. The alloy structure consists of a Y-matrix and a finely dispersed g/-phase that is uniformly distributed in it. As a result of high rates of abrasive-erosive wear, the alloy undergoes structural transformations, with coagulation, changes in the morphology of the strengthening Y/-phase and carbides in the body and along the grain boundaries. Further operation of RDP is impossible, although the resource of their work is not exhausted [92 – 94].

The main disadvantage of the mining complex is a decrease in the physical and mechanical properties of the surface and corrosion destruction of the structure, as well as high roughness, which, when the oil fluid interacts with the released acid-salt radicals of the metal, degrades the structure of the RDP. A.G. Gazarov, L.A. Nagirny, T.V. Prikhodko high mechanical properties of nickel alloy products achieved by electrolytic deposition technology [95, 96]. The main problem is the need to control the microstructure, porosity and compliance with the balance of tensile and compressive stresses. Consequently, in the CIS, the technologies of designing and manufacturing mining complexes lag far behind modern realities. An actual method is manufacturing using highly concentrated sources of laser-plasma energy.

The studies of Ya.I. Frenkel, V.N. Dovbysh, other scientists and the authors of the project showed that laser technologies allow controlling the granularity of the microstructure and controlling the martensitic and austenitic medium of the material, modifying surfaces, and not limiting the stresses in the details, but using their effect (compression–stretching) [97 – 102]. Scientific and technological needs are concentrated in the absence of energy-efficient technology for laser modification of the inner surface of a small diameter. An important problem that needs to be solved is not only the development of a unique technology for restoring the structure and properties of the pump metal, but also to develop an original installation that ensures the restoration of the inner surface of the pump with a diameter of up to 44 – 50 mm over the entire length of the pump 4 – 6 m. Existing analogues provide hardening of the contact surface, but do not have technical solutions for creating and modifying the structure of the base material and the design geometry of the plunger parts of the complex with its partial loss. Thus, the fundamental difference of the research idea lies in the development and justification of a unique technology for laser modification of the structure and design geometry of RDP with specified physical and mechanical properties.

The motive for the development of innovative recovery technology was the shortcomings of the method of manufacturing a deep pumping complex of low-flow wells. The market of production and restoration services is poor in the supply of high-quality manufacturing of mining complexes due to strict requirements.

The scientific problem is to substantiate the optimal material with high adhesive and corrosion-wear properties, providing phase modification of the coating depending on the modes of laser modification.

Scientists S.B. Beketov, V.N. Arbuzov, L.V. Ivanova, E.A. Burov, M.L. Galimullin, D.A. Shock, J.O. Sudbury, J.J. Crockett proved that deposits in different wells differ from each other in chemical composition depending on the group hydrocarbon composition of different types of oil [53, 103 – 107].

The authors of the research have established the scientific novelty of the interaction with certain types of metals, quartz compounds and zirconium dioxide. Paraffin molecules participate in co-crystallization with alkyl chains of asphaltenes, forming a point structure. As a result, paraffin does not form a solid lattice, being redistributed between many small centers and the release of paraffins on the surface is significantly weakened.

Conclusion

The solution to the problem of cylinder wear is to cover the inner surface of the hole with a ceramic protective layer, high hardness and inert properties. Plasma and laser spraying technologies are used for spraying ceramic powder. The small diameter (Ø38 – Ø44 mm) limits the use of plasma spraying technology. The dimensions of the plasma spraying unit are 1.5 – 3 times larger than the possible installation based on laser spraying technology.

When using laser spraying technology, the lenses for focusing the laser beam and the light guides can be positioned separately from the feeding device of the deposited material (powder) or in some cases outside the working device. Also, during plasma spraying, in a limited space, the gas jet affects a large area (compared to a laser beam), as a result of which it is possible to melt the sprayed layer, its separation from the part and a change in the relief of the surface layer, which worsens its strength. In turn, the laser beam acts pointwise on the surfacing site and does not cause overheating of the part with deterioration of strength characteristics.

From the analysis of technologies, it can be seen that for surfacing in conditions of limited space, laser spraying is more suitable for use, since it has fewer critical disadvantages compared to plasma spraying. Modern laser installations exceed the efficiency of plasma installations in terms of weight and dimensions, and unit costs are also reduced due to the low energy intensity of the technology and high efficiency.

An actual solution is to harden the working surface of the cylinder with ceramic materials based on zirconium (ZrO_2 – zirconium dioxide). Zirconium dioxide has a wide range of useful properties (Table 8) [108, 109]:

- high corrosion resistance;
- high crack resistance among ceramic materials;
- low thermal conductivity;
- maintaining strength over a wide temperature range;
- antifriction properties;
- inertia to oil production products.

Mechanical processing in order to bring to the exact dimensions (grinding, honing) of metal-ceramic alloys, does not differ from the processing of ordinary metals and alloys.

Table 8. Ceramic alloys based on zirconium dioxide [109]

Properties	Metal-ceramic alloys			
Composition	$ZrO_2 + Y_2O_3$	$ZrO_2 + MgO$	$ZrO_2 + CaO$	$ZrO_2 + Al_2O_3$
Density, [g/cm³]	5.8-6.05	5.6-5.7	5.6-5.7	5.4-5.6
Open porosity, [%]	0	0	0	0
Hardness, [HRC]	72	72	72	72
Flexural strength, [MPa]	300 – 1,000	500 – 600	500 – 600	1,900 – 2,100
Compressive strength, [MPa]	2,000 – 2,200	1,800 – 1,900	1,800 – 1,900	1,900 – 2,100
Thermal conductivity at 20 – 100 °C, Вт/мК [W/mK]	2.0 – 2.5	2.0 – 2.5	2.0 – 2.5	5 – 7
Coefficient of linear thermal expansion at 20 – 1000 °C	10 – 11	10 – 11	10 – 11	5 – 7
Maximum operating temperature, [°C]	1,000	1,000	1,000	1,000

The most effective and modern method for applying zirconium dioxide to metal parts is the spraying technology using concentrated light sources. Laser technology allows for an ultrashort period of time, on a small area of the surface, to obtain high temperatures (more than 3,000 °C) sufficient for melting zirconium dioxide ($T_{melting}$ = 2,715 °C) [109].

There is a need to study the technologies of spraying powder materials on the surface of small diameter holes, as well as the technology of moving devices inside deep holes. It is necessary to determine the optimal design parameters that allow the use of laser technologies to harden internal surfaces of small diameter, as well as to develop a design of a laser device for free movement inside the cylinder of a rod pump and the application of coatings that change the physical and mechanical properties of the surface.

For the successful implementation of laser recovery technology, it is necessary to develop an original design of an adaptive installation for restoring the physical and mechanical properties of metal products of the inner surface of a small diameter pump.

Chapter 3. Development of production-technological equipment for special purposes

3.1 Analysis of existing technological equipment

The search and analysis of technological equipment for pretreatment of RDP cylinders has shown that domestic and foreign analogues of pre-boring equipment are divided into several main types of construction (Table 9).

Table 9. Types of boring tools

Type of boring head	Advantages	Disadvantages	Operating modes
1. Single-blade boring head with internal supply of lubricating-cooling fluid (LCF) and two-row combined guide elements.	1. Double-row guides. 2. The first row is made in the form of rigid fixed slats. 3. The second row consists of nylon slats in grooves, under which plates of elastic material are added to increase the shock-absorbing properties (polyurethane, hard rubber).	1. One blade limits performance. 2. Soldered tool. 3. The difficulty of adjusting the bore diameter (an additional device is needed).	Rough boring of holes from 50 mm to 250 mm: - revolutions – 350 rpm; - tool feed – 60 –100 mm/min.
2. The boring head of Botek (Germany) of one-sided cutting with the certainty of basing for boring deep holes.	1. The certainty of basing ensures high centering accuracy. 2. The possibility of quick replacement of carbide plates. 3. Nylon guides are used to dampen vibrations.	1. Adaptations are needed to adjust the bore diameter. 2. Low performance.	Rough boring of holes from 45 mm to 250 mm: - revolutions – 450 rpm; - tool feed – 60 – 90 mm/min.
3. Boring heads of Sandvik Coromant (Sweden) are single-cut with a certain basing with ejector chip removal.	1. When drilling holes from 40 mm to 180 mm, the carbide plates are installed in the block to fine-tune the boring diameter. 2. Ejector chip removal allows it to be removed through the channel of the connecting stem after the drilling operation.	1. For boring holes from 20 mm to 40 mm, the groove for the plate is made in the housing without an adjustment block. 2. Ejector chip removal complicates the design of the stem and increases the requirements for equipment.	Rough boring: - revolutions – 400 rpm; - tool feed – 70 –100 mm/min.

| 4. Two-blade boring head of double-sided cutting with the division of the thickness of the cut and limited movable guides. | 1. The plates are arranged diametrically opposite, which increases the processing performance. 2. The movement of the movable guides is limited due to the constant pressing against the base surface, which ensures high quality centering of the block. | Orientation of the boring head to rough boring. | Rough boring: - revolutions – 400 rpm; - tool feed – 150 mm/min. Finishing boring: - revolutions – 750 rpm; - tool feed – 250 mm/min. |

In single-blade boring heads of the first type, a significant disadvantage is a cutter with a soldered plate. When replacing the cutting tool, it is necessary to re-manufacture the cutter and subsequently this may affect the cutting parameters (accuracy, cutting speed (V), roughness (Ra)) since it is impossible to create a geometrically identical cutter. The disadvantage is also the need for a device to control the departure of the tool from the body of the boring head. As a result, the required technological parameters are achieved by trial and error, which increases the complexity of the work and reduces the quality of the surface treatment of the part. This technical problem can be solved by replacing the soldered tool with a prefabricated one. This will reduce the labor costs for changeover and the technological process itself.

The next disadvantage of a single-blade tool is the unidirectionality of the cutting forces P_y (Figure 43). The force P_y is directed towards the relative axis of rotation of the tool and tends to push (displace) the cutting plate in the radial direction.

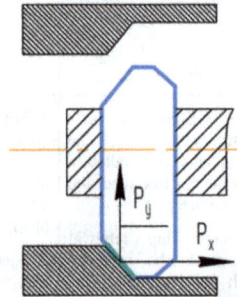

Figure 43. Axial (P_x) and radial (P_y) cutting forces

This property affects the increase in the loads experienced by the guides located opposite to the cutting edge of the tool. In this connection, the wear intensity of the guides occurs unevenly and the axis of symmetry of the centering of the head is shifted.

Sandvik heads with ejector chip removal are recommended to be used after drilling the workpiece. At the same time, there is no need to reconfigure chip removal systems and lubricating-coolant liquid (LCL) on a drilling-boring machine, which saves auxiliary time.

The two-blade boring head of double-sided cutting with the division of the cut thickness allows you to get the tool away from the axis within 0.5 mm/pog.m. The design implies the use of a large number of precision parts that complicate the adjustment of the boring head. To reduce the cost of production and operation of the head, it is recommended to replace the dependent release mechanism using cones with independent guides in which the task of the release mechanism is performed by inserts made of elastic materials.

To improve the centering efficiency of the boring head, reduce the cutting resistance forces and ensure the alignment of the axes of symmetry of the cutting tool and the axis of the head, it is proposed to develop an original design of a two-blade boring head with the alignment of the tool block with a pin. It is proposed to use independent guides made of carbide plates as centering elements.

In order to ensure the optimal mode of chip removal from the cutting zone and to prevent the guides from entering the area with simultaneous lubrication, the location of the liquid outlet holes is taken out of the guides closer to the connection of the stem with the boring head. The supply of LCF is carried out through the stem, and the chips are diverted in the direction of the tool feed.

3.2   Calculation of cutting modes satisfying the effective operation of the developed boring head

In order to increase the efficiency of the technological process of processing the inner surface of the cylinder RDP, it is necessary to calculate the optimal cutting modes. The practical problem is that the classical calculation methods are presented for ideal conditions (factory setting of the machine, tool sharpening, and so on). However, operational factors of tool wear, technical condition of equipment, deviation from the axis of impact on long cylinders are not taken into account. To solve this problem, it is necessary to make a preliminary calculation of the initial parameters under the given conditions and assumptions. Further, on the basis of the obtained areas of optimal values, to adapt to specific production conditions. For the preliminary calculation, a feed equal to S=0.25 mm/rev is accepted.

The cutting depth (t) is calculated by the formula (23):

$$t = \frac{D-d}{2},$$ (23)

where D – the required diameter, mm;

d – preliminary diameter, mm.

$$t = \frac{37.64 - 32}{2} = 2.82 \;\; mm.$$

Purpose of the cutting speed V:

$$V = \frac{C_v}{T^m t^x S^y} \cdot K_v \cdot 0,9,$$ (24)

where C_v – speed coefficient when boring, C_v=420;

T – tool durability period, T=15 min;

t – cutting depth, mm;

S – supply, mm/rev;

x, y, m – degree indicators that take into account the influence of processing

modes, x=0.15, y=0.20, m=0.20.

The product of coefficients (Kv) taking into account the influence of cutting factors is calculated by the formula (25). The calculation results are presented in Table 10.

$$K_v = Kmv \cdot Kpv \cdot Ktv \cdot K\varphi v,$$ (25)

where K_{mv} – a coefficient that takes into account the quality of the processed material, K_{mv}= 0.77;

K_{pv} – a coefficient reflecting the condition of the work piece surface,

$K_{pv}= 0.9$;

K_{tv} – a coefficient that takes into account the quality of the tool material, $K_{tv}= 0.9$;

$K_{\varphi v}$ – a coefficient that takes into account the main angle in the plan (φ, °).

Table 10. Kv at the main angle in the plan (φ, °)

The main angle in the plan is φ, [°]	$K_{\varphi v}$	K_v
20	1.4	0.87
30	1.2	0.75
45	1	0.62
60	0.9	0.56
75	0.8	0.5
90	0,7	0.44

We determine the cutting speed taking into account the feed S, the tool life reserve T, and the value of the removable allowance t on the side. The calculation results are presented in Table 11.

Table 11. Influence of angle φ on cutting speed V

The main angle in the plan is φ, [°]	V, [m/min]
20	218
30	187
45	155
60	140
75	125
90	110

The calculation of the spindle speed (n) for boring the inner diameter of the cylinder RDP, with a single-slot design of the boring block, is determined by the formula (26), rpm:

$$n = \frac{1{,}000 \cdot V}{\pi \cdot D},$$
(26)

where 1,000 – conversion coefficient of millimeters to meters;

V – cutting speed, m/min;

D – the inner diameter of the cylinder RDP, mm.

After calculating the revolutions at a given cutting speed, it is necessary to bring them in line with the equipment used for deep boring (Figure 44). The assigned actual spindle revolutions (n_{act}) and cutting speed V_{act} are shown in Table 12.

Figure 44. Machine spindle speed panel PT2632

Table 12. Assigned actual spindle revolutions (n_{act}) and cutting speed V_{act}

V, [m/min]	n, [rpm]	n_{act}, [rpm]	V_{act}, [m/min]
218	1,844	1,600	189
187	1,582	1,500	177
155	1,311	1,300	154
140	1,184	1,100	130
125	1,057	1,000	118
110	930	900	106

The actual revolutions are limited to the maximum supported revolutions of the machine spindle $n_{max} = 1{,}600$ rpm.

In the machine, the minute feed of the V_s tool is used to adjust the feed, mm/min. The minute feed indicates how far the tool will travel in a certain amount of time. Recalculation of the S feed into the minute V_s feed makes it easier to set up the machine later. To convert the feed value per revolution into a tool feed per minute, it is required to make calculations according to the formula (27). The correspondence of the minute feed to the actual revolutions is shown in Table 13.

$$V_s = S \cdot n_{act} , \qquad (27)$$

where V_s – minute tool supply, mm/min;

S – supply, mm/rev;

n_{act} – actual spindle revolutions, rpm.

Table 13. The correspondence of the minute feed to the actual revolutions

n_{act}, [rpm]	V_s, [mm/min]
1,600	400
1,500	375
1,300	325
1,100	275
1,000	250
900	225

To increase the processing speed and reduce the influence of the cutting force, a constructive decision was made to use two incisor plates by placing them diametrically opposite. To ensure technologically specified boring modes, with the use of a new design of the boring head, it is required to increase the feed S twice. The increase in feed is due to the fact that it is necessary to maintain the constancy of the removable allowance for one plate. The increase in feed has a direct proportional effect on the minute cutting speed V_s (mm/min). After finding the V_s value, it is necessary to round the values to the nearest machine-supported V_{cs} (Figure 45).

Figure 45. Machine feed panel PT2632

It is necessary to double the feed to ensure optimal operation of both edges of the plate and maintain the cutting speed. The actual minute feed rate ($V_{act\,s}$) is calculated using the formula (27). The calculation results are presented in Table 14.

For optimization and efficiency, an algorithm of the calculation methodology using the Walter Machining Calculator software application is proposed. The results of calculations of cutting power and torque are presented in Table 15.

Table 14. Correspondence of tool feed and machine feed

n_{act}, [rpm]	$V_{act\,s}$, [mm/min]	V_{cs}, [mm/min]
1,600	800	854
1,500	750	693
1,300	650	693
1,100	550	562
1,000	500	456
900	450	456

Table 15. Cutting power and torque

The main angle in the plan is φ, [°]	Torque, M_t, [N/m]	Power, N, [kW]
20	62.57	12.33
30	56.15	10.37
45	62.52	10.00
60	60.55	8.20
75	55.61	6.85
90	60.18	6.6

The dependence graph in Figure 46 shows how the power consumed by the machine equipment varies depending on the angle φ,°.

From the dependence graph (Figure 46) it can be seen that with an increase in the main angle in plane of φ, the power consumed by the machine to perform the cutting process decreases, and the equipment experiences significantly less loads on the drive structural elements. However, reducing the load on the machine drive by increasing the main angle in terms of φ does not guarantee high surface treatment efficiency. Therefore, in order to perform the boring of the RDP with high efficiency, it is necessary to justify the optimal values of the main angle in plane of φ. To do this, it is necessary to conduct additional research.

In order to select the optimal cutting parameters, it is necessary to calculate the forces acting on the cutting tool. The required cutting forces allows us to investigate the design for the possibility of using cutting modes in which these forces arise.

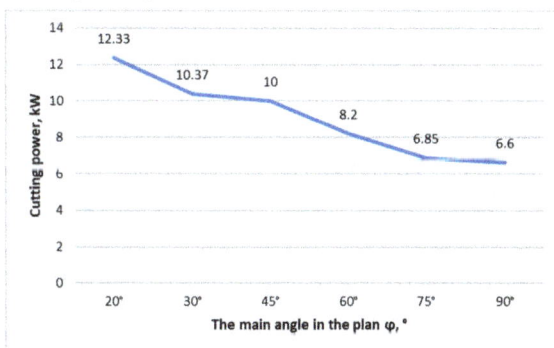

Figure 46. Dependence of the change in cutting power N on the main angle in plane of φ°

To calculate the cutting force P_z, the inverse method is used using the formula for calculating the cutting power. The P_z force acts tangentially to the circumference of the cylinder bore and is directed perpendicular to the cutting plane at each point. P_z denotes the force experienced by the tool during machining. With a known cutting power, the force P_z per plate is calculated by the formula (28). The calculation results are presented in Table 16.

$$P_z = \frac{1{,}020 \cdot 60 \cdot N}{2 \cdot V}, \tag{28}$$

where P_z – cutting force per plate, N;

$\quad$ N – cutting power, kW;

$\quad$ V – actual cutting speed, m/min.

Table 16. Final cutting parameters

The main angle in the plan is φ, [°]	V_{act}, [m/min]	Power, N, [kW]	P_z, [N]
20	189	16.24	1,996
30	177	10.37	1,792
45	154	10.00	1,987
60	130	8.20	1,930
75	118	6.85	1,776
90	106	6.6	1,905

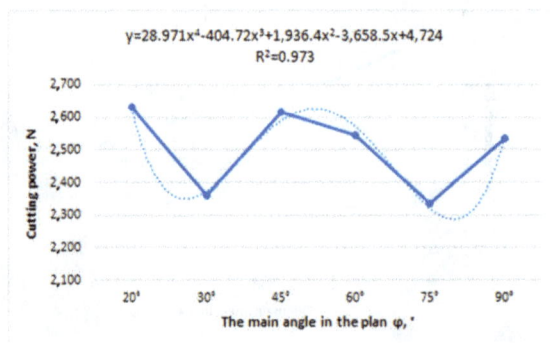

Figure 47. Change of the cutting force P_z from the main angle in the plan φ

Analyzing the dependence (Figure 47), it was found that at angles of 30° and 75°, a decrease in the cutting force P_z is observed. Based on these data, it can be concluded that angles of 30° and 75° are the most suitable for the design of the boring block. The reduction of the cutting force P_z is due to the fact that the machine does not allow you to choose the optimal feed equal to the calculated one and there is a need to choose values close to the calculated values.

For the verification calculation of the dependence of the change in the P_{zv} force, static variables are needed, such as the cutting speed and the spindle speed of the machine. With these unchangeable indicators, it is possible to plot the influence of the main angle in the plan on the cutting force. To calculate, we select the revolutions that ensure maximum stability of the machine system with the part (vibrations from the rotation of the part, minimum cutting force). From the calculated data, the best option is the turns n = 1,000 rpm and the cutting speed V_{act} = 118 m/min. In this case, according to the formula (7) of the test calculation of the cutting force:

$$P_{zv} = 10C_{pz}t^x S^y V_{act}^n K_{pz} \qquad (29)$$

where P_{zv} – cutting force during the test calculation, N;

C_{pz} – the constant component of the axial force, C_{pz} = 300;

t – cutting depth, mm;

S – supply, mm/rev;

V_{act} – actual cutting speed, m/min.

x, y, n – degree indicators that take into account the influence of processing modes: $x = 1, y = 0.75, n = -0.15$;

K_{pz} – the coefficient that takes into account the geometry of the tool, calculated by the formula (30). The calculation results are presented in Table 17.

$$K_{pz} = K_{mpz} \cdot K_{\gamma pz} \cdot K_{\lambda pz} \cdot K_{\varphi pz} \qquad (30)$$

where K_{mpz} – a coefficient that takes into account the quality of the processed material, K_{mpz} = 1.05;

$K_{\gamma pz}$ – coefficient depending on the front angle $\gamma°$, $K_{\gamma pz}$ = 1;

$K_{\lambda pz}$ – coefficient depending on the angle of inclination of the plate, $K_{\lambda pz}$=1;

$K_{\varphi pz}$ – a coefficient that takes into account the main angle in the plan (φ, °).

Table 17. K_{pz} at the main angle in the plan (φ, °)

The main angle in the plan is φ, [°]	$K_{\varphi pz}$	K_{pz}
20	1.20	1.26
30	1.08	1.12
45	1.00	1.05
60	0.94	1.00
75	0.91	0.96
90	0.89	0.93

The results of calculating the dependence of the test cutting force P_{zv} on the angle φ are presented in Table 18.

Table 18. Dependence of the test cutting force P_{zv} on the angle φ

The main angle in the plan is φ, [°]	P_{zv}, [N]
20	1,828
30	1,625
45	1,523
60	1,436
75	1,392
90	1,349

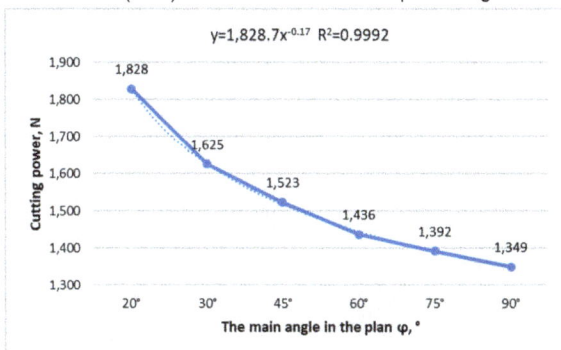

Figure 48. Graph of the dependence of the cutting force on the main angle in the plan

An increase in radial force is unacceptable in an insufficiently rigid machine system, namely, with a large reach of the boring head (up to 6 m), the rigidity of the stem does not participate in centering, and the task is completely transferred to the basing guides. In this case, the resulting increase in the P_y force creates the risk of the tool moving away from the axis with inaccurate installation of the cutting plates relative to each other and with a change in the internal diameter and shape of the surface.

Based on the above, reducing the radial force in a tool based on the principle of feed division is an important task in ensuring high-quality hole processing. To calculate the radial force P_y, the formula is used:

$$P_y = 10C_{py}t^x S^y V_{act}^n K_{py} \tag{31}$$

where C_{py} – the constant component of the radial force, $C_{py} = 243$;

t – cutting depth, mm;

S – supply, mm/rev;

V_{act} – actual cutting speed действительная скорость резания, m/min;

x, y, n – degree indicators that take into account the influence of processing modes; $x = 0.9$, $y = 0.6$, $n = -0.3$;

K_{py} – a coefficient that takes into account the geometry of the tool and is calculated by the formula (32):

$$K_{py} = K_{mpy} \cdot K_{\gamma py} \cdot K_{\lambda py} \cdot K_{\varphi py} \qquad (32)$$

where K_{mpy} – a coefficient that takes into account the quality of the processed material, $K_{myp} = 1.05$;

$K_{\gamma py}$ – coefficient depending on the front angle $\gamma°$, $K_{\gamma py} = 1$;

$K_{\lambda py}$ – coefficient depending on the angle of inclination of the plate, $K_{\lambda py} = 1$ [15, p. 275, *Table 23*];

$K_{\varphi py}$ – the coefficient taking into account the main angle in the plan, $\varphi°$ is given in Table 19.

Table 19. K_{py} at the main angle in the plan (φ, °)

The main angle in the plan is φ, [°]	$K_{\varphi py}$	K_{py}	P_y, [N]
20	1.50	1.575	1,003
30	1.30	1.365	869
45	1.00	1.050	668
60	0.77	0.810	515
75	0.61	0.640	407
90	0.50	0.520	331

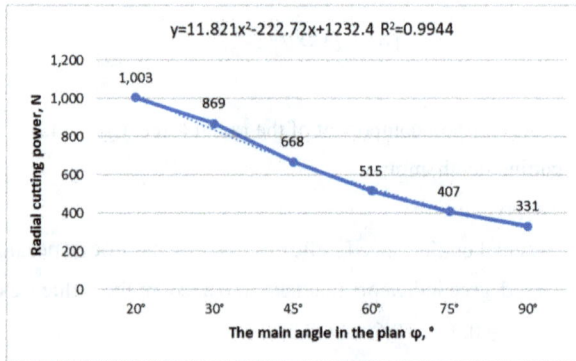

Figure 49. Dependence of the radial cutting force on the main angle in the plan

From the indicators of the graphs (Figure 49) it can be seen that the optimal design for the boring rough block is the main angle in the plan equal to 75°. This angle provides an optimal cutting speed of 118 m/min while maintaining the maximum effective forces (P_z and P_y) acting on the boring head and affecting the quality of the bore hole. The angle of 30° is less suitable for use in the construction of the boring block, since with a decrease in the angle φ, the risk of parasitic vibrations increases. This is due to the concomitant increase in the radial cutting force (P_y), which, together with a decrease in chip thickness, a large contact line length and low rigidity of the machine system, causes uneven cutting.

3.3 Strength calculation of the body of the developed boring head

To obtain optimal values of the boring head design (Figure 50), it is necessary to conduct simulation modeling. The study of strength characteristics was carried out in the APM-FEM software environment.

The strength calculation of the boring head was performed in the КОМПАС-3D modeling program using the APM-FEM module. For the initial calculation, the minimum value of the cutting force at 75° was selected, $P_{zv} = 1,625$ N per plate.

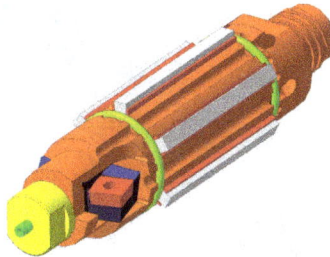

Figure 50. General view of the boring head

To reduce the load on the computer, elements (shank, front part with thread) that do not affect the final strength analysis were excluded from the calculation by means of a cross section. Loading was carried out on a segment of the face of the plate equal to the cut-off allowance.

According to the color palette (Figure 51), it can be seen that the voltage concentration in the body of the housing and the tool block does not exceed the permissible, as evidenced by the green and blue colors of the spectrum corresponding to $\sigma_u = 800$ MPa for 40Cr Steel. In the area of the plate, namely the area interacting directly with the metal, a force of 1,210 MPa acts. Further, all stresses are distributed evenly on the housing and there are no foci of metal overvoltage due to the high strength and rigidity of the boring head design. The strength condition corresponds to the value of the bending strength of the carbide plate $\sigma_F = 2,500$ MPa.

The minimum value of the fluidity safety factor is observed under the plate on the tool block (Figure 51).

Name	Type	Minimum value	Maximum value
Equivalent Mises voltage	SVM [MPa]	3.566833	1,209.720252

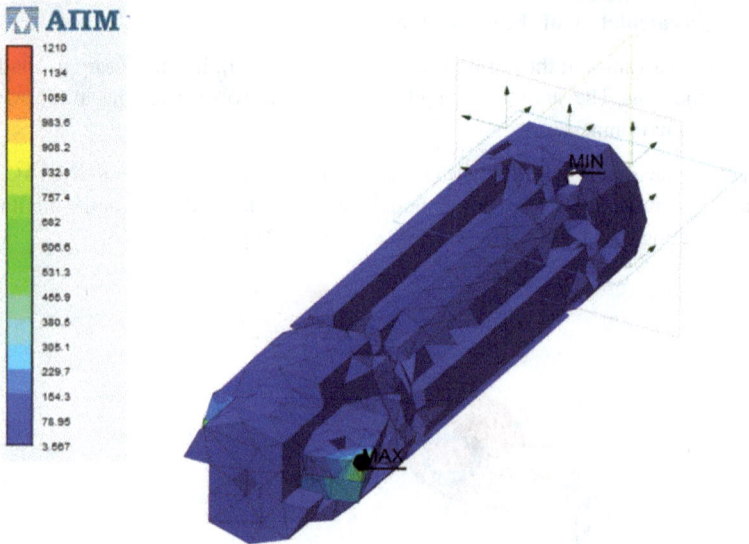

Figure 51. Results of static stress calculation

The margin factor (Figure 52) at the minimum point equal to 3.227 satisfies the requirement for maintaining operability (more than 2), but for guaranteed preservation of operability, it is recommended to increase it to a value above 4. For this, it is necessary to reduce the cutting force P_z. It is possible to reduce the force by reducing the minute feed of the tool, as a result of which the size of the cut layer decreases in one revolution of the spindle. Reducing the speed and, as a consequence, the cutting speed will not give a positive result, since reducing the cutting speed negatively affects the surface quality (*Ra* and Rz) and the resulting uniformity of the metal cutting process, especially in a non-rigid system at long processing lengths.

The safety factor for strength (Figure 53) meets the requirements of normal working capacity. The critical stress points, in this case, are located in the area of the tool block under the plates. The body is painted in orange shades, which corresponds to the safety factors from 51.69 to 99.35. These values are many times higher than the value of the safety factor (K > 4).

Name	Type	Minimum value	Maximum value
A safety factor for fluidity		3.226522	1,000

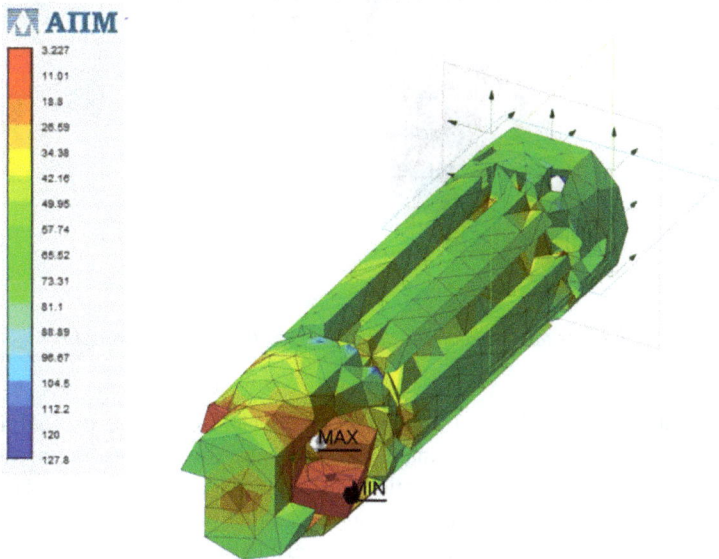

Figure 52. Results of static calculation of the yield strength factor

Name	Type	Minimum value	Maximum value
A safety factor for strength		4.028015	1,000

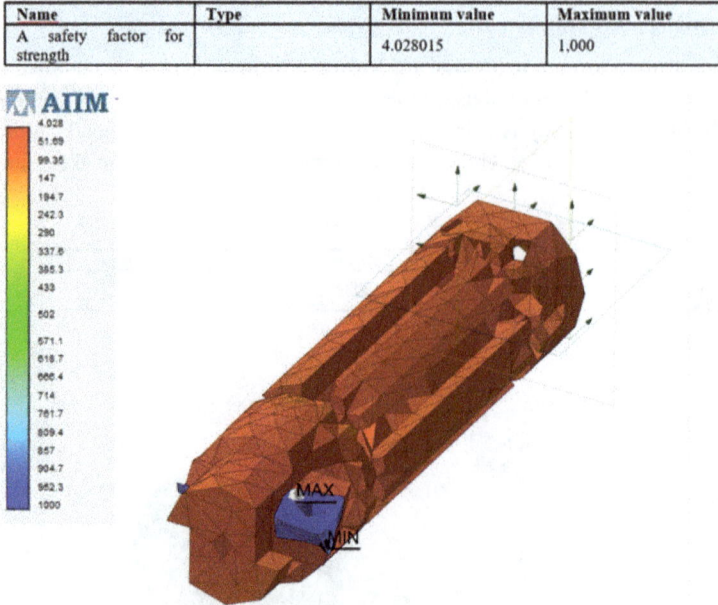

Figure 53. Results of static calculation of the safety factor for strength

During the operation of the boring tool working on the principle of dividing the feed (two cutters), the probability of vibrations is high. The occurrence of vibrations is associated with the low rigidity of the Machine-Adaptation-Tool-Part (MATP) system during boring of cylinders of small diameters (20 mm – 44 mm). Vibrations have a detrimental effect on the surface quality during operation, tool wear, structural failure, microcracks under the influence of vibrations increase the chance of destruction of the tool structure. In this regard, it is necessary to carry out a vibration calculation of the structure to identify the destructive vibration effect, which can cause undamped vibrations (self-oscillations), with prolonged exposure to which the structure loses its strength characteristics. The results of the natural frequencies of the boring head are presented in Table 20.

Table 20. Report of the results of the natural frequencies of the boring head

№	Frequency, [rad/sec]	Frequency, [Hz]
1	35.114072	5.588575
2	37.428891	5.956993
3	40.87756	6.505866
4	60.168085	9.576048
5	60.168111	9.576050

Analyzing the tabular values of the frequency characteristics of the part (Table 20), we can conclude that the frequency of 5 Hz occurs during processing with low cutting speeds (from 10 to 50 m/min). The phenomenon of increasing vibrations is associated with the chip formation process, the geometry of the plates. Low cutting speed provokes not cutting metal with a cutter, but squeezing it with the edges of the plate and, as a result, its accelerated wear. To solve this problem, it is necessary to prevent chip formation with a frequency equal to the natural oscillation frequency of the boring head and increase the cutting speed.

An increase in the cutting speed gives an effect with an advanced geometry of the chip breaker groove (Figure 54). Carbide plates with a similar chip breaker design are characterized by an increased price (from 2 to 10 times more expensive) compared to plates with standard (simple) chip-breaking grooves. The plate shown below has a price 5 times higher compared to the plate of a simplified design (Figure 55).

Figure 54. A plate with a complex geometry of the chip-breaking groove

The simple geometry of the chip-breaking groove (Figure 55) is more economically acceptable to use, but requires refinement with the chip-breaking and chip-removal mechanism.

Figure 55. Plate with simplified geometry of the chip-breaking groove

Simple geometric parameters of the chip breaker do not guarantee the prevention of the formation of drain chips, which is unacceptable during deep boring. The drain chips, passing through the hole in the cylinder, with the flow of liquid accumulates inside the workpiece and forms a jam, after which further processing of the workpiece is impossible.

The solution to the problem is the geometry of the developed boring block design, which combines the calculated optimal main angle in terms of φ and the cutting angle γ, which ensure continuous and stable formation of element chips.

Conclusion

The design of the boring head for cutting threaded grooves in the cylinder bore has been developed. The technology involves the cutting of a torn thread, and therefore the calculation of the optimal parameters of the cutting angles for maximum efficiency of the boring process was carried out. It was found that at angles of 30° and 75°, a decrease in the cutting force P_z is observed. Based on these data, it can be concluded that angles of 30° and 75° are the most suitable for the design of the boring block. However, due to the low longitudinal rigidity of the system, an increase in radial force is unacceptable. In the insufficiently rigid system of the machine, namely, with a large reach of the boring head (up to 6 m), the rigidity of the stem does not participate in centering, and the task is completely transferred to the basing guides. In this case, the resulting increase in the P_y force creates the risk of the tool moving away from the axis with inaccurate installation of the cutting plates relative to each other, as well as with uneven wear of the plates. Based on the above, reducing the radial force in the tool is a priority, and therefore the main angle in the plan (φ) was chosen equal to 75°, providing an optimal cutting speed of 118 m/min.

Chapter 4. Controlled installation for laser spraying of internal surfaces of small diameter

4.1 Description of a controlled installation for laser spraying of internal surfaces of small diameter

Multifunctional mobile laser unit   (Figure 56), includes a laser radiation source, an optical system of mirrors, including a lens, a deflecting module and a protective glass installed in the original spraying unit, consisting of a nozzle with a mechanism for adjusting the angle of powder and laser beam (Figure 57, position 1), a carrier mobile housing of cylindrical execution of the spraying process, a feeding carriage with powder composition feeding channels, which provides axial stabilization of the spraying unit inside the cylinder bore for a continuous and uniform spraying process (Figure 57, position 2), an adjustment slider with a tilt angle control motor mounted on it, a linear drive system for moving adjustment elements, sliding and gear engagement hinge blocks, a tapered nozzle with an adjustable diameter for feeding powder material, roller converters of rotational motion into translational motion, a mechanism for stabilizing the axial position of the spraying unit relative to a rotating pipe with a compensating cylinder with axial rollers, fixed by a clamping bar, a gear train in the form of controlled satellites with a flat gear of internal and end engagement, a module for placing optical fiber and a cable channel for electrical harnesses.

Figure 56. Automated unit of laser spraying

In addition, the universal nozzle for feeding the powder composition is made of two components. Its lower part has the shape of a cone with a stepped base for fixing, a cylindrical central hole provides the axial direction of the laser, confusor-diffusor grooves are made along the outer cone surface to supply powder material with increased pressure. The second part of the nozzle is made in the form of a conical lid and is screwed onto the lower part, the landing depth regulates the cross-sectional area of the opening hole. The unification of the laser installation and its modularity of systems ensures adaptive multi-angle movement of the running carriage and the accuracy of automated positioning of the laser head in adaptive mode to the speed and power parameters of the spraying process.

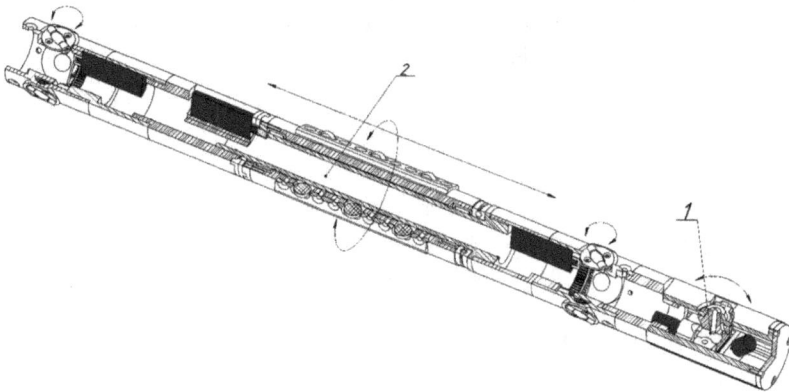

Figure 57. Main parts of the unit

The possibility of achieving the technical result of processing long pipes of small diameter is ensured by the fact that the body of the spraying head moves independently in the pipe opening, using the torque of the rotating pipe as the driving force, the gear mechanism and the roller hinges of the mirror holder platform ensure the adaptation of the laser reflection angle depending on the movement, and the supply of the sprayed material is carried out with a flow protective gas through the confusor-diffuser holes.

The system of linear electric drives provides automated control of each module and individual elements, adapting to high-speed and power modes of movement. The clamping bar with axial rollers ensures effective pressing against the inner surface and moving the installation along the workpiece. The feeding mechanism with a gear planetary gear makes it possible to change the angle of rotation of the feeding roller and automatically change the direction of movement of the laser installation. Adaptive stabilization of the laser nozzle relative to the position of the axis of the rotating part is provided by an axial sleeve and a compensation cylinder rotating on its axis, which is pressed against the surface of the part by axial rollers and controls the speed. The optical fiber passed through the spraying head is attached to the spraying unit and outputs the laser beam to the mirror, which has automatic adjustment for a given spraying angle. The proposed optical fiber layout eliminates the need for more mirrors, thereby increasing the efficiency of the laser. Optical fiber allows you to achieve the minimum dimensions of the installation, as well as the length of the optical fiber does not limit the depth of movement of the laser spraying head inside the treated cylinder.

Laser blocks connected to optical fiber have a modular structure, which allows, if necessary, their replacement, disassembly of the installation for transportation, simplified repair without affecting the main parts of the spraying head.

The declared set of constructive and technological solutions and the conditions for the implementation of these actions to achieve the effect are not explicitly identified from the known state of the art and technology.

Consequently, these features of the claimed invention have not been identified in other technical solutions, which means that the solution is new, which meets the criteria of patentability "novelty" and has an inventive level.

4.2 Components of the developed spray head

A controlled laser installation for the restoration and spraying of the inner surface of small diameter long pumps (Figure 58) contains a laser radiation source (not shown in the figure).

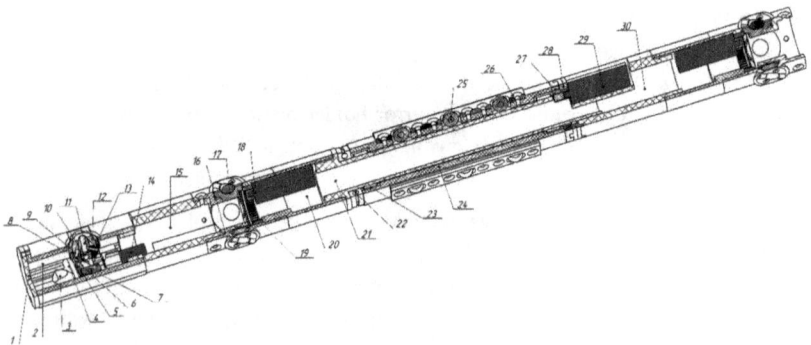

Figure 58. Laser spraying head assembly

The essence of the model is explained by the following description and drawings:

1. general view of the laser spraying head of internal surfaces of small diameter (Figure 58);
2. spraying node (Figure 59);
3. feeding node (Figure 60);
4. the stabilization mechanism (Figure 61).

The laser spraying head (Figure 58) in its design contains an end cap 1 mounted on the body of the laser head 2, in which the tilt angle motor 3 is placed on the adjusting slider 4, a coupled toothed hinge 5 with a mirror 6 through the hinge 7. The design also contains a nozzle housing 8, which is the basic part for installing a pressure hood 9 for fixing the protective glass 10, a focusing tip 11 and an adjustment cone 12 with a cone mount 13. A linear actuator 14 serves as a regulator of the angle of inclination of the nozzle apparatus.

Next, a coupling 15 is installed, which is a transitional link for coupling with the housing of the feeding mechanism 16 from the feeding unit (Figure 60), on which a guide drive roller 17 is mounted, driven by a feed motor 18 through the engagement of a gear train of a flat gear 19. The coupling of the feeding unit (Figure 60) with the stabilization mechanism (Figure 61) is carried out thanks to the transition coupling 20 connected to the axial sleeve 21. The mechanism also contains thrust 22 and needle 23 bearings, a compensation cylinder 24, axial rollers 25 with a clamping bar 26, an internal gear 27, a nylon bearing housing 28 and a compensation cylinder motor 29 rigidly fixed in the compensator housing 30.

The retention of the spraying unit (Figure 59) and the feeding unit (Figure 60) from simultaneous rotation together with the workpiece is provided by a stabilization mechanism (Figure 61). The stabilization of the position is carried out by reverse rotation of the compensation cylinder 24, relative to the rotation of the workpiece, through the gear engagement of the internal gear 27 with the gear of the engine of the compensation cylinder 29.

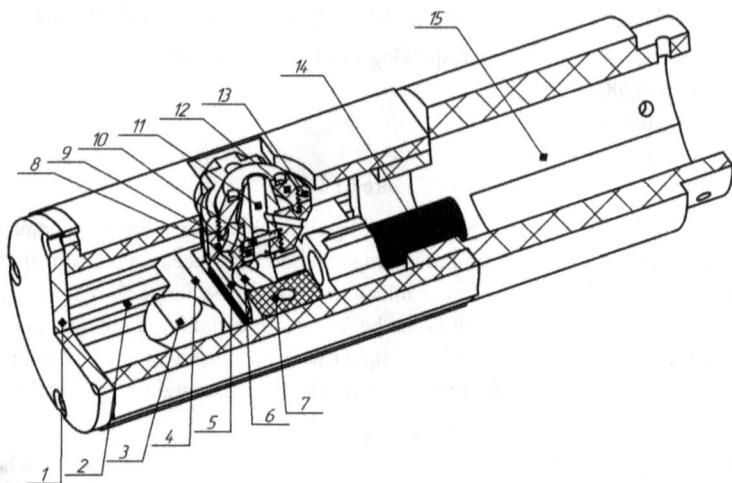

Figure 59. Spraying node

The entire laser spraying unit is controlled automatically after the operator sets the input parameters of the feed and the angle of inclination of the nozzle. The angle of reflection of the mirror 6 is interdependent on the position of the adjustment slider 4 and is a multiple. Thus, when moving the linear actuator 14 within 8 mm, it is possible to set the spray angle to 30°. The nozzle angle is set due to the swivel connection through the formed lever.

All drives of the spraying unit (linear drive 14 and tilt angle motor 3) are controlled synchronously and interdependently according to the coordinates set in the program.

The laser beam passes through the optical fiber (not specified), to the spraying node (Figure 59) heading towards the mirror 6 and, depending on the angular position of the toothed hinge 5, reflecting passes through the hole of the nozzle body 8, protected from ingress of the sprayed material by a protective glass 10, which is fixed by a pressure hood 9 and passes through the hole of the focusing tip 10. The angle of reflection of the mirror 6 is adjusted through the rotation of the toothed hinge 5 by rotating the motor shaft of the tilt angle 3.

The sprayed material is transported by process gas through the confusor-diffuser openings of the nozzle body 8 and the adjusting cone 12 acts as the focusing element of the flow, which, due to the threaded connection with the attachment of the cone 13, converts rotational motion into translational motion, which makes it possible to establish an output gap between the focusing tip 11 and the adjusting cone 12. To set the spraying angle, the adjusting slider 4 is moved to a predetermined distance by a linear actuator 14 and through the hinge 7, the nozzle body 8 and the entire nozzle apparatus consisting of a focusing tip 11 and an adjusting cone 12 are rotated.

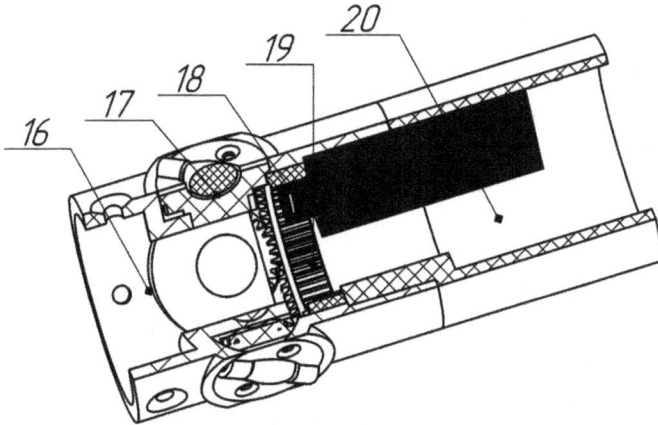

Figure 60. Feeding node

The transformation of the rotational motion of the workpiece into the translational motion of the laser head is provided by the feeding unit (Figure 60). To move at the required speed (mm/rev), an angle is set on the guide drive roller of the feed 17 by rotating the shaft of the feed motor 18 through the gear drive of the flat gear 19. Thus, setting the angle on the guide drive roller of the feed 17 with relative rotational, opposite movements of the rotating workpiece with a fixed feeding unit, a spiral is formed along which the surface of the drive roller 17 is rolled and moves the body of the laser head in a linear direction along the axis of rotation.

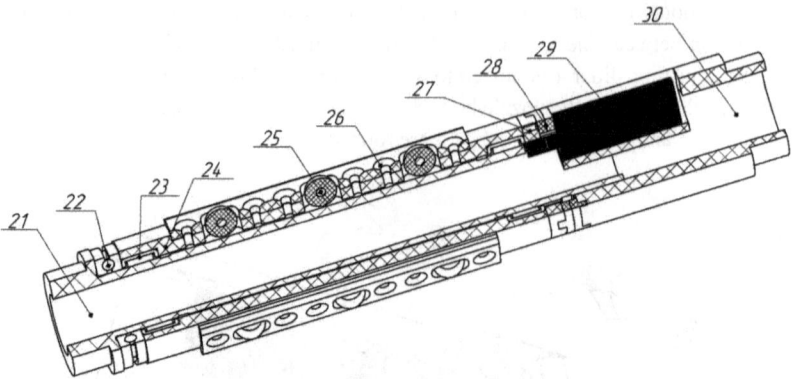

Figure 61. Stabilization mechanism

The axial sleeve 21 is the basic part of the stabilization mechanism, on which the thrust 22 and needle 23 bearings are placed with a compensation cylinder 24 fixed to their rotating support with axial rollers 25, which are fixed by a clamping bar 26 and have one degree of freedom of rotation in the axial direction. The axial rollers 25 are pressed against the surface of the workpiece to exclude the rotation of the compensation cylinder 24. An internal gear 27 is attached to the compensation cylinder 24. The rotation is facilitated by the nylon bearing of the housing 28, which also fixes the adjacent parts from axial movement. The motor of the compensation cylinder 29, rigidly fixed in the compensator housing 30, which is connected to the main part of the laser installation, gives a rotational moment to the entire stabilization mechanism (Figure 61).

4.3 Development of a management system

Ensuring the necessary characteristics of the developed device is achieved by using a control system based on an Arduino microcontroller in the design. The microcontroller allows you to apply a modular structure in the design of the spraying head and expand the functionality of the tasks performed. The modular structure has high manufacturability and maintainability and, in combination with additive technologies, allows for high-speed repair, replacement and modernization of the spray head.

The design of the spray head uses such electronic components as: electric motors with a gearbox, stepper motors, rotation angle sensors, a speed control sensor, an accelerometer module with a gyroscope.

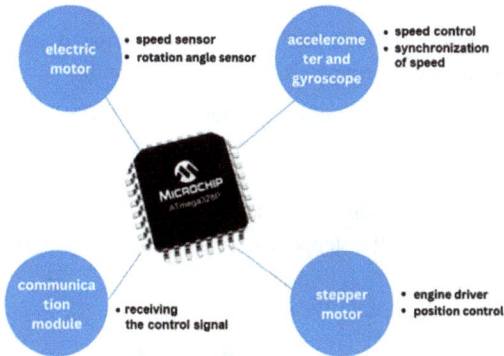

Figure 62. Modules of the control system of the spraying unit

The diagram (Figure 62) shows the main and auxiliary modules that perform various functions to ensure the operability of the laser unit. The ATmega328P microcontroller is the primary information processing module that receives the received information from the communication module based on the NRF24L01+ radio module. On the basis of the ATmega328P microcontroller [110] (Figure 62), a fully functional miniature Arduino Nano device [111] is built, to the terminals of which all the above modules are quickly connected.

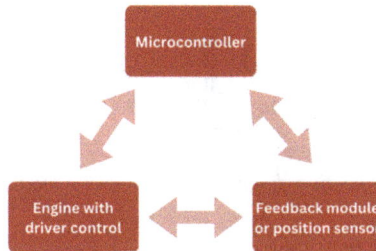

Figure 63. Simplified scheme of the control system device

Figure 63 shows a simplified diagram of the control system of the laser spraying head. The scheme implies that the main control function is performed by the microcontroller, processing the readings from the sensors and applying them to calculate the signal to the motor drivers, which set the exact movement or rotation of the working body (motor).

Tracking and transmitting information about the operation is performed by feedback modules that are mechanically connected to the working body.

Controlled operation of stepper motors and DC motors is carried out through control drivers, which receive a control signal from a microcontroller and process it, driving the motor. A stepper motor can perform precise movements or make an exact number of revolutions without feedback. This result is achieved by setting in the control program the required number of steps that the stepper motor must take (Figure 64, 65).

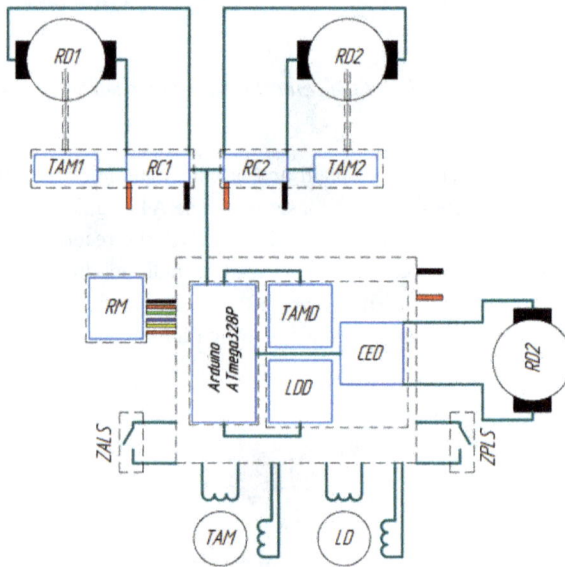

Figure 64. Connection diagram of the working modules of the unit

To position the feed drive roller (Figure 60, position 17) using a DC motor with a planetary gearbox, which is installed in the laser head in the feeding node (Figure 60, position 18), a 5 kOhm potentiometer with a 180° rotation angle is used. The potentiometer performs the functions of a rotation angle sensor due to changes in the voltage at the signal output from the angle and the conversion of its value by a digital-to-analog converter into digital values for further calculation by the microcontroller.

Also, a DC motor with a gearbox is used to control the stabilization of the installation (Figure 61, position 29). Rotation control takes place in a constant mode and is carried out using a sensor of the number of revolutions made on the basis of an optocoupler. The rotation speed of the compensation cylinder (Figure 61, position 24) must coincide with the rotation speed of the cylinder being processed with high accuracy, which does not allow maintaining the stability of the nozzle position in the vertical direction.

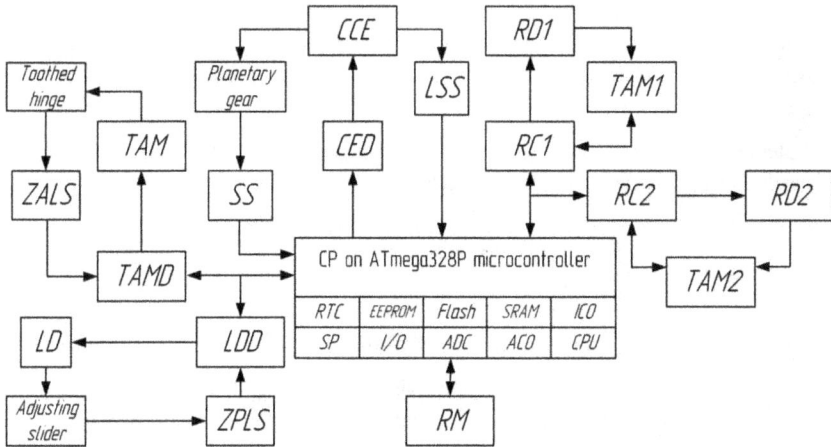

Figure 65. The scheme of interaction of the unit modules

Table 21. Designation of modules in the diagram

Spraying node	LDD	Linear drive driver (драйвер)
	LD	Linear drive
	ZPLS	Zero position limit switch of the linear actuator
	TAMD	Tilt angle motor driver
	TAM	Tilt angle motor
	ZALS	Zero angle limit switch
Feed node	RC1...2	Drive roller rotation controller
	RD1...2	Rotation driver
	TAM1...2	Turning angle motor
Stabilization node	CED	Compensation engine driver
	CCE	Compensation cylinder engine
	SS	Rotation speed sensor of the compensation cylinder
	LSS	Limit switch stop of the extreme position

To correct the nozzle position error from the vertical, an additional adjustment sensor was introduced, combining an accelerometer and an MPU5060 gyroscope in one chip. The gyroscope determines the physical position of the laser installation in space, and the correction of displacement, sudden movements and jerks is made taking into account the measurements of the accelerometer. The calculated values of MPU5060 are applied to the specified speed of rotation of the cylinder and when skewed from the vertical, towards the direction of rotation, the correction value is subtracted from the specified speed of the cylinder, when skewed in the opposite direction from rotation, the correction is added. Thus, by changing the rotation speed of the compensation cylinder, the vertical of the nozzle of the spraying head is corrected.

Two stepper motors are installed in the spraying node (Figure 59), one of which has a screw shaft and a guide in the design for linear movement in the range from 0 to 8 mm and a stepper motor with a gear on the shaft that engages with the teeth of the gear joint (Figure 59, position 5) and regulates rotation the angle of the mirror (Figure 59, position 6).

4.4   Calculation of optimal characteristics of the laser source

Spraying is a very complex process that requires fine-tuning of technological equipment to obtain a high quality coating. One of the main characteristics that ensures the quality of spraying is the choice of a laser source.

To calculate the characteristics of the laser source, it is necessary to determine the optimal flow rate of the gas-powder mixture through the nozzle. For the calculation, it is necessary to determine the throughput of the nozzle section, the amount of sprayed material and the required gas flow rate for transfer.

The power, area and processing speed directly affect the possibility of effective activation of the melting process of the sprayed material and transfer to the surface. The dependence of the effect of laser radiation on the speed of movement of the laser point and the power of the laser source are presented in the formula [112]:

$$T \approx \frac{2}{\sqrt{\pi}} \frac{q_0(1 - R)}{k} \sqrt{\frac{2ar}{V_p}} + T_i, \tag{33}$$

where q_0 – radiation power density [113], W/cm^2;

R – surface reflection coefficient, for zirconium dioxide $R = 0.7$;

k – thermal conductivity of zirconium dioxide, $k = 3.5$ W/m·K;

a – the coefficient of thermal conductivity of the material, mm/sec^2;

r – radius of the beam impact area, r = 0.5 mm;

V_p – the speed of the powder granules (Figure 66);

T_i – initial temperature, $T_i = 22$ °C.

The thermal conductivity of a substance is calculated by the formula:

$$a = \frac{k}{\rho_z \cdot C} \tag{34}$$

where k – thermal conductivity of zirconium dioxide, $k = 3.5$ W/m·K; (вставила)

ρ_z – density of zirconium dioxide, $\rho_z = 5.68$ g/cm^3;

C – specific heat capacity [114], $C = 0.45$ J/g·K.

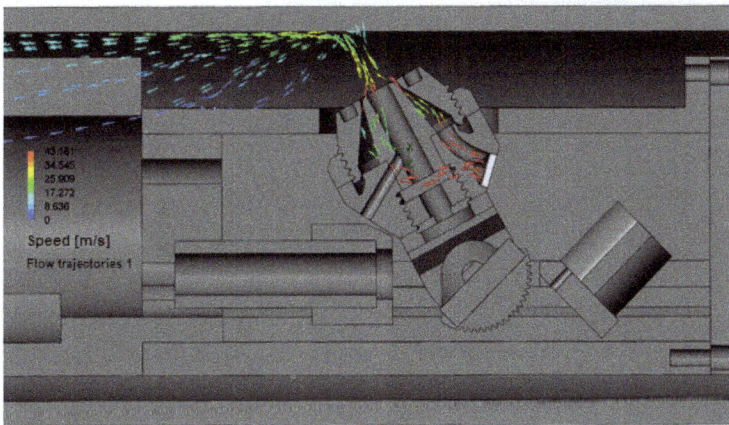

Figure 66. Simulation of the flow rate of the protective gas with particles of the sprayed powder

The simulation allowed us to prove that with a mass flow rate of *1 gram/s*, the velocity of the sprayed particles in the convergence zone of the flow at the surface is $V_f = 17$ m/s.

Figure 67 shows a linear increase in density from the power of the laser source, but a nonlinear change from the radius of the impact area. To identify the optimal range of radii and laser powers, it is necessary to calculate the temperatures that occur at the specified parameters.

The temperature in the zone must correspond to the temperature necessary for the activation of the melting process of zirconium dioxide in the range from the melting temperature of 2,715 °C to the boiling point of 4,300 °C. The calculation will determine the required maximum and minimum power of the laser unit, the choice of the bandwidth of the fiber-optic cable, the need for cooling the fiber-optic cable.

The radiation power density (q_0) implies what power is concentrated on the area illuminated by the laser. The radiation power density depends on the radius of the beam area (r) and the power of the beam emitted by the laser. Figure 68 shows the change in the radiation power density.

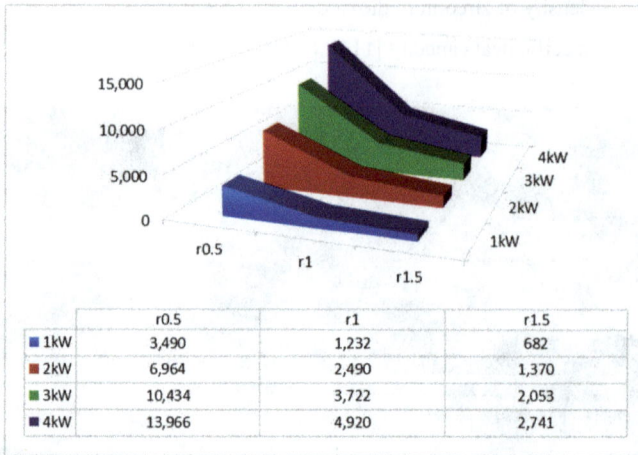

	r0.5	r1	r1.5
■ 1kW	3,490	1,232	682
■ 2kW	6,964	2,490	1,370
■ 3kW	10,434	3,722	2,053
■ 4kW	13,966	4,920	2,741

Figure 67. Temperature distribution in the beam area of lasers of different power

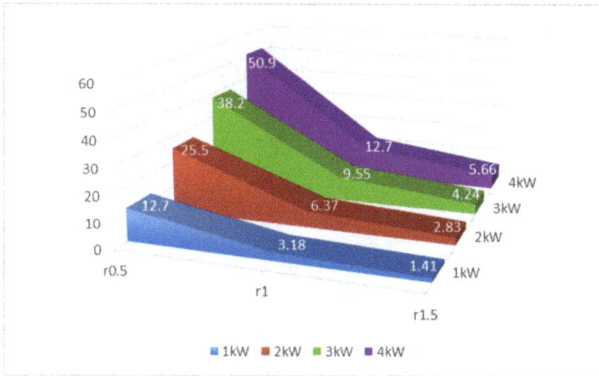

Figure 68. Dependence of the power density on the radius of the impact area and laser power, 10^4

The fiber-optic cable affects the overall characteristics of the laser head because, when the laser power is more than 1.5 kW, liquid cooling must be used for the focusing head (Figure 69).

It is proved that in order to activate the melting process of particles in the convergence zone, the optimal parameters of the power and radius of the illumination area are:

– for a radius r = 0.5 mm, it is optimal to use a laser beam with a power of 1 kW;
– for a radius r = 1 mm, it is optimal to use a laser beam with a power from 2.5 kW to 3.5 kW.

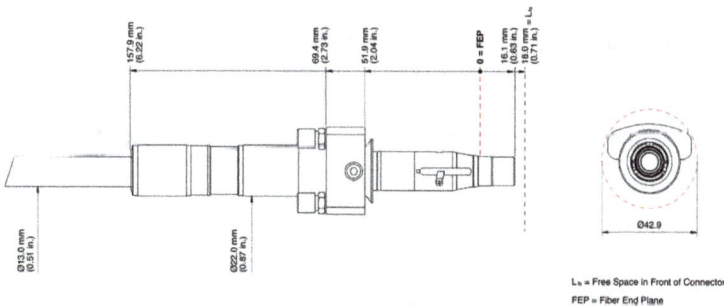

Figure 69. High-power optical fiber, with a capacity of 15 kW, equipped with a liquid-cooled connector [115]

It is recommended to use a laser unit with a power of 1 kW and a COHERENT RQB Fiber Optic Cable (Figure 70).

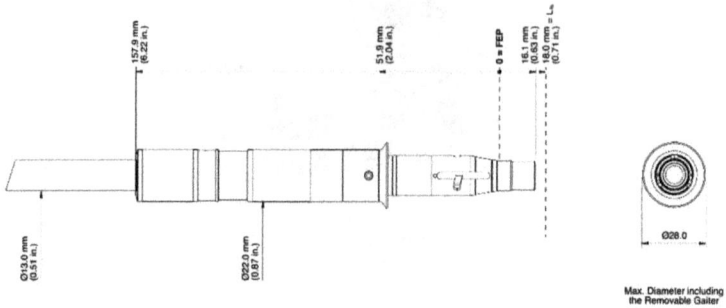

Figure 70. Drawing of the output connector of the fiber optic cable [115]

The compactness of the connector without water cooling allows you to freely place the optical system in the laser head. A laser beam with a diameter of 1 mm is directed at the reflecting mirror and passes through the protective mirror to the area of convergence with the flow of the sprayed substance. The diameter of the laser beam of 1 mm provides a fiber core with an optical connector system with a diameter of 1 mm.

To optimize the deposition process, it is necessary to calculate the rate of deposition of the material on the cylinder surface. Due to the fact that the possibility of the powder melting process in the illumination area is precisely determined, it is necessary to calculate the throughput of the nozzle apparatus.

It is determined that a layer with a thickness of at least $S_l = 0.6$ mm must be sprayed on one revolution of the cylinder, thus, using the density of the sprayed substance, it is possible to calculate the mass of the site. During one revolution, the installation is guaranteed to spray substances of at least 0.48 grams of the substance, or 0.0846 g/cm³ in volume.

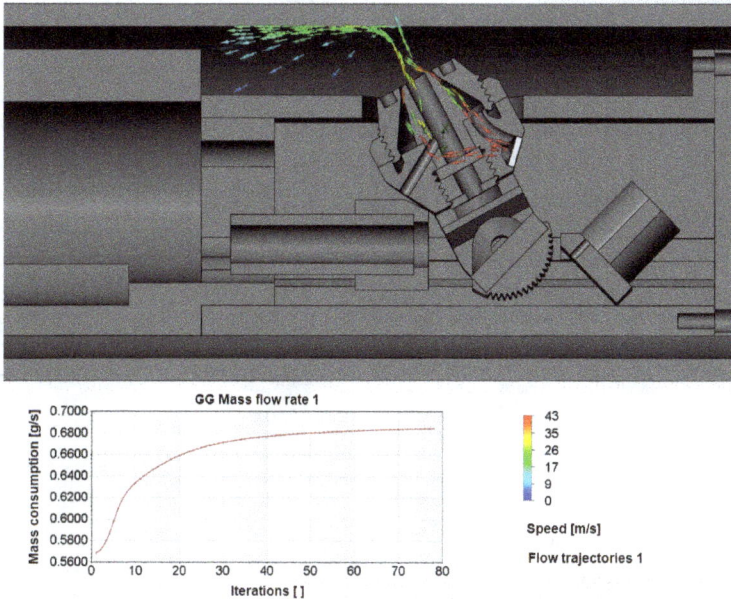

Figure 71. Software calculation of the mass flow rate of gas in FloEFD

Computer modeling (Figure 71) in the FloEFD program proved that the mass flow rate of gas, for the transfer of 0.48 grams of powder into the melt zone at a speed of 17 m/s, must correspond to 0.7 grams/s. Accordingly, the feed of the laser head in the sprayed cylinder is defined as the product of time per stroke and is equal to 60 mm/min. The flow capacity of the nozzle allows you to achieve a set value due to the adjustable gap between the flow section formed by the surfaces of the nozzle cone and the adjusting cone. The threaded connection with a step of 1 mm allows you to fine-tune the flow of the gas-powder mixture depending on the size of the granules used and the gas flow rate.

4.5   Development of technology for manufacturing a rod pump cylinder using a laser spraying unit

An important technological task for any type of spraying is to ensure high quality adhesion of the sprayed material to the workpiece [116]. Contamination and oils are not allowed on the sprayed surface of the part, which lead to a deterioration in the adhesion of the sprayed material to the substrate. The most common methods of surface preparation for spraying are divided into two types – mechanical and chemical methods [117]. Most often, mechanical methods of preparation are used, such as shot processing, knurling, sand blowing, slicing, and so on [118]. The most effective method is threading with subsequent processing with a fraction. The adhesion strength during such processing is calculated by the tensile strength and is equal to 190 MPa, but there is a risk of reducing the fatigue and strength of the part [119]. In this case, the cylinder does not experience alternating loads that can lead to the destruction of the part. Therefore, it was decided to apply the method of threading with a boring head followed by shot blasting (Table 22).

Table 22. The route of laser treatment of the RDP cylinder

005	Install the workpiece on the device for editing.	
Procurement operation	Measure the straightness to match 0.1 by 1 meter. If the straightness does not match, edit the workpiece.	
010	Grind the outer diameter to the full length, withstand the size Ø57.$_{-1}$	
Centerless grinding operation	Maintain a tolerance of roundness within 0.1 mm.	

015	Process the part according to the control program, maintaining the dimensions: 1, 2, 3, 4, 5, 6, 7.	
Turning NCS (numerical control software)	Withstand roughness Ra1.6. Maintain a perpendicular tolerance of 0.02 mm. Withstand angles of 45°, 30°, 15°.	
020	Bore the inner diameter to a size of Ø45.5±0.03 with the formation of a 0.25 mm high thread.	
Boring operation	To withstand the deviation of straightness in the range of 0.1 by 1 meter. Maintain the alignment tolerance within 0.4 mm.	
025	Process the inner surface of the cylinder with a fraction for the entire length.	
Shot blasting operation		

030 Turning NCS	Install the part with the untreated end face and set the runout relative to A no more than 0.1 mm. Sharpening dimensions 1, 2, 3, 4, 5, 6, 7 according to the outline in the sketch. Turn the part 180°, install, fix. Waste size 8 according to the outline on the sketch.	
035 Spraying	Install the spray head inside the cylinder. Spray a layer of ceramics up to a diameter of $\varnothing44.3^{+0.15}$.	
040 Honing operation	Install the part, set the radial runout of 0.1 mm according to the indicator. Honing size $\varnothing44.45^{+0.05}$.	

Conclusion

In order to simplify the production and expand the technological possibilities of hardening the inner surface of the RDP cylinder, an automated installation for laser spraying of powdered materials was developed. The installation is capable of hardening the surface layer in holes of small diameter (44 mm) and restoring worn coatings in places of greatest wear. The patent search conducted showed the absence of analogues of this device and the uniqueness of the design solutions adopted.

The prototype of an automated laser spraying installation is made by 3D prototyping and implies the use of 3D printing technology in an industrial sample. With 3D printing technology, it is possible to produce parts of the installation that are not exposed to high temperatures, which simplifies the production of the laser head. On the manufactured prototype, a system of movement and positioning inside the RDP cylinder has been worked out.

Conclusion

Analytical and experimental studies have been carried out in the field of laser restoration of the inner surface of small-diameter mining pumps and the design of an automated installation for laser spraying of powdered materials has been developed.

It has been experimentally established that paraffin deposits and mechanical impurities are distributed unevenly along the diameter of the working cylinder of the discharge column, narrowing the flow section of the rod in different stroke length intervals. The difference in the through diameter in the pipe causes the rod to deviate from the design axis of the trajectory, creating a pendulum stroke and shock loads. Such a process of operation of the pump leads to an uneven distribution of friction forces and stress concentration, which accelerates the mechanical wear of the hardened layer of the inner surface, reducing the operability ("cylinder-plunger" pair) of the pump.

An actual solution for the restoration of the inner surface of small–diameter mining pumps is an energy-efficient technology for modifying the working surface of the cylinder with ceramic materials based on zirconium (ZrO_2 - zirconium dioxide), which has a wide range of useful properties.

In order to expand the technological possibilities of hardening the inner surface of the cylinder of the RDP, an automated installation for laser spraying of powdered materials has been developed. The installation is capable of hardening the surface layer in holes of small diameter (Ø38 – 44 mm), up to 6 meters long and restoring worn areas of the inner surface of the product in places of greatest wear.

The conducted patent search showed the absence of analogues of this device and the uniqueness of the design solutions adopted. The automated installation of laser spraying is carried out by 3D prototyping and involves the use of 3D printing technology in an industrial sample. With 3D printing technology, it is possible to produce parts of the installation that are not exposed to high temperatures, which simplifies the production of the laser head. On the manufactured prototype, a system of movement and positioning inside the RDP cylinder has been worked out.

Additionally, due to the specifics of the production of the cylinder, the design of the boring head for cutting threaded grooves in the cylinder bore has been developed. It has been experimentally proved that with the main angle in the plan (φ) equal to 75°, there is a decrease in the cutting force and radial force in the tool, which is a priority indicator of the reliability of this system.

Materials Research Forum LLC
https://doi.org/10.21741/9781644902356

References

[1]   Effective solutions for the operation of a low-capacity fund of oil wells. Triol Corporation.   13.12.2016.   URL: https://triolcorp.ru/news/post/effektivnye-resheniya-ekspluatatsii-malodebitnogo-fonda-neftyanykh-skvazhin

[2]   Beisekov S.S., Kurbanov R.R., Balgynova A.M. (2016) Increasing the coefficient of oil recovery of viscous oils // PETROLEUM. No. 6(102). URL: https://www.petroleumjournal.kz//index.php?p=article&aid1=79&aid2=403&id=9 64&outlang=1

[3]   Utemisova L.G., Tlegenov B.B., Minikaev F.M. (2021) Efficiency analysis of simultaneous and separate operation of wells at the Zhetybai field // Bulletin of the oil and gas industry of Kazakhstan. No. 1(6). pp. 67-74. URL: https://vestnik-ngo.kz/2707-4226/article/download/88897/67075

[4]   Ivanova T.N., Novokshonov D.N., Galeeva O.A., Bartashova A. (2020) Analysis of the effectiveness of the used installations of rod depth pumps in conditions of high-viscosity oil production // Bulatovskie readings. V2. pp. 218-224. URL: https://elibrary.ru/item.asp?id=43810396

[5]   Frolov A.A., Gluhenky A.M. (2019) Analysis of the work of the mechanized well fund for the first half of 2019 of the NCCDPNG NGDU "Ufaneft" and the development of measures to increase the OHP and SNO GNO. Report of NGDU UFANEFT LLC. – 63 p. URL: https://ppt-online.org/622622

[6]   Well operation by RDP units. The community "ANO DPO OTSTPK and NTI". 17.02.2018.   URL: https://vk.com/@dopprofobraz-ekspluataciya-skvazhin-ustanovkami-shgn

[7]   Goldobin V. (2010) Time of plunger pumps // Oil of Russia. – No. 6. – pp. 68-70.

[8]   Brunman V.E., Vataev A.S., Volkov A.N. et al. (2017) Energy-efficient oil extraction by sucker rod borehole pumps in low-yield fields. Russ. Engin. Res. 37, 378 382. URL: https://doi.org/10.3103/S1068798X17050082

[9]   Brunman V.E., Vataev A.S., Volkov A.N. et al. (2017) Improving the energy efficiency of borehole pumps. Russ. Engin. Res. 37, 579-580. URL: https://doi.org/10.3103/S1068798X17070061

[10]   Shcherba V.E., Bolshtyanskii A.P., Kaigorodov S.Y. et al. (2016) Benefits of integrating displacement pumps and compressors. Russ. Engin. Res. 36, 174-178. URL: https://doi.org/10.3103/S1068798X1603014X

[11]   Liu H., Yan J., Meng S. et al. (2019) Practice and understanding of developing new technologies and equipment for green and low-carbon production of oilfields. Front. Eng. Manag. 6, 517-523. URL: https://doi.org/10.1007/s42524-019-0061-0

[12]   Petrushin E.O., Arutyunyan A.S. (2015) Hydrodynamics' studies of the bore holes on formed mode // Scientific journal "Postgraduate Student". Mining. Development of oil and gas fields. No. 4(9). – pp. 179-184. eLIBRARY ID: 24094218. URL: https://www.elibrary.ru/item.asp?id=24094218

[13]   Shagiev R.G. (1988) Investigation of wells by KVD (pressure recovery curves). – Moscow: Nauka, Moscow. – 304 p.

[14]   Asalkhuzina G.F., Davletbaev A.Ya., Abdullin R.I., Gareev R.R., Shiman A.P., Loshak A.A., Filev M.O. (2021) Welltesting for a linear development system in low permeability formation. Petroleum engineering. Oil and Gas Fields Development. V. 19, № 3. – pp. 49-58. https://doi.org/10.17122/ngdelo-2021-3-49-58

[15]   Alekseev A.D., Aniskin A.A., Volokitin Ya.E., Zhitny M.S., Karnauh D.A., Khabarov A.V. (2011) Experience and prospects of application of modern GIS and GDIS complexes at the fields of the Salym group // Industrial and technical oil and gas journal "Engineering Practice" (issue 11-12'2011).

[16]   Features of equipment and technology of oil production by ECPP units // oilloot.ru.   URL: http://oilloot.ru/80-dobycha-i-promyslovaya-podgotovka-nefti/510-osobennosti-tekhniki-i-tekhnologii-dobychi-nefti-ustanovkami-uetsn

[17]   Sherstyuk A.N., Trulev A.V., Ermolaeva T.A. et al. (2003) Features of the Characteristics of Submersible Centrifugal Oil Pumps. Chemical and Petroleum Engineering 39, 23-26. URL: https://doi.org/10.1023/A:1023782205956

[18]   Valyukhov S.G., Zhitenev A.I. & Zhitenev S.V. (2009) Modern pumps made by Turbonasos Enterprise for pumping oil and oil products. Chem Petrol Eng 45, 144-147. URL: https://doi.org/10.1007/s10556-009-9162-7

[19]   General scheme of installation of the electric centrifugal pump // agrovodcom.ru. URL: http://www.agrovodcom.ru/infos/uetsn-ustanovka.php

[20]   GOST 32601-2013 (ISO 13709:2009, MOD) Centrifugal pumps for the oil, petrochemical and gas industries. General technical requirements. The publication is official. Moscow: Standartinform, 2015. – 307 p.

[21]   Centrifugal   pump   (ECP)   //   Rengm.ru.   URL: https://rengm.ru/rengm/centrobezhnyy-nasos-ecn.html

[22] Overview of the technology of oil pumps ECP // agrovodcom.ru. URL: http://www.agrovodcom.ru/info_obzor_ezn.php

[23] Mittag S., Gabi M. (2015) Experimental and numerical investigation of centrifugal pumps with asymmetric inflow conditions. J. Therm. Sci. 24, 516–525. URL: https://doi.org/10.1007/s11630-015-0817-8

[24] Ivanovsky V.N., Degovtsov A.V., Sabirov A.A., Krivenkov S.V. (2017) Influence on the operating time of installations of electric centrifugal pumps of supply and pump rotation speed during the operation of wells complicated by removal of mechanical impurities // Territory of Neftegaz. – No. 9 – pp. 58-64

[25] Markelov D.V. (2000) Operating experience with Russian and imported immersed centrifugal pumps at the Yugankneftegas company. Chem Petrol Eng 36, 157-160. URL: https://doi.org/10.1007/BF02463414

[26] Tong Zm., Xin Jg., Tong Sg. et al. (2020) Internal flow structure, fault detection, and performance optimization of centrifugal pumps. J. Zhejiang Univ. Sci. A 21, 85-117. URL: https://doi.org/10.1631/jzus.A1900608

[27] Fakher S., Khlaifat A., Hossain M.E. et al. (2021) Rigorous review of electrical submersible pump failure mechanisms and their mitigation measures. J Petrol Explor Prod Technol 11, 3799-3814. URL: https://doi.org/10.1007/s13202-021-01271-6

[28] Rod depth pumps: design, principle of operation, varieties // rosprombur.ru. URL: https://rosprombur.ru/shtangovye-glubinnye-nasosy-konstrukciya-princip-raboty-raznovidnosti.html

[29] Purpose, design and technical characteristics of the rocking machine // stanokgid.ru URL: https://stanokgid.ru/specializirovannyj/stanki-kachalki.htmlnokgid.ru

[30] Shiyan S.I., Sleptsov A.A., Sukhoverkhova P.A. (2020) Analysis of production capabilities and technological modes of wells equipped with PRDP at the Sasimov oil field // Branch scientific and applied research: Earth Sciences. – No. 4. – pp. 228-242.

[31] GOST R 31825-2012 (2013) Downhole rod pumps. General technical requirements. Official publication. Moscow: Standartinform, p. 48. URL: https://files.stroyinf.ru/Data/537/53731.pdf

[32] Nasretdinov M.R. Self-installing magnetic valve of the barbed deep pump. Patent RF, RU 185543 U1. Byull. No. 34. 10 December 2018. URL:

https://patents.google.com/patent/RU185543U1

[33] Shkandratov V.V., Bortnikov A.E., Sidorov D.A., Kazakov A.A., Astafyev D.A. Oil-well sucker-rod pump. Patent RF, RU 2321772 C1, Byul. No. 10, 10 April 2008. URL: https://patents.google.com/patent/RU2321772C1

[34] Stepanov N.V., Efimov A.A. Borehole rod pump. Patent RF, RU 96617 U1, Bul. No. 22, 10 August. 2010 URL: https://patents.google.com/patent/RU96617U1

[35] Vedeneev A.I. Borehole plug-in sucker rod pump. Patent RF, RU 163755 U1, Byul.   No.   22,   10   August   2016.   URL: https://patents.google.com/patent/RU163755U1

[36] Urazakov K.R., Seitpagambetov Zh.S., Valeev M.D., Kutluyarov Yu.Kh. Sucker-rod pumping plant. Patent RF, RU 2175402 C1, 27 October 2001. URL: https://patents.google.com/patent/RU2175402C1

[37]   Pyalchenkov D.V. (2016) Investigation of the influence of the parameters of producing wells on the failures of rod pumping units // Online journal "Science Studies" V. 8, No. 2. pp. 1-10. DOI: 10.15862/140TVN. URL: http://naukovedenie.ru/PDF/140TVN216.pdf

[38] Ivanova T.N., Emel'yanov E.O., Novokshonov D.N. et al. (2016) Study of Pump Efficiency and Causes of Pump Failure. Chem Petrol Eng 52, 344-346. URL: https://doi.org/10.1007/s10556-016-0197-2

[39] Gupta J., Kujur A., Prasad S.R. et al. (2016) Analysis of Frequent Failure of Oil Well Tubings Due to Formation of Longitudinal Slit. J Fail. Anal. and Preven. 16, 518-526. URL: https://doi.org/10.1007/s11668-016-0120-3

[40] Repair of wells equipped with rod borehole pumps // neftegaz.ru URL: https://neftegaz.ru/tech-library/burovye-ustanovki-i-ikh-uzly/141496-remont-skvazhin-oborudovannykh-shtangovymi-skvazhinnymi-nasosami/

[41] Fakher S., Khlaifat A., Hossain M.E. et al. (2021) A comprehensive review of sucker rod pumps' components, diagnostics, mathematical models, and common failures and mitigations. J Petrol Explor Prod Technol 11, 3815-3839. URL: https://doi.org/10.1007/s13202-021-01270-7

[42] Abdelazim A., Abu El Ela M., El-Banbi A. et al. (2022) Successful approach to mitigate the asphaltenes precipitation problems in ESP oil wells. J Petrol Explor Prod Technol 12, 725-741. URL: https://doi.org/10.1007/s13202-021-01335-7

[43] Al-Qasim A., Al-Anazi A., Omar A.B., Ghamdi M. (2018) Asphaltene precipitation: A review on remediation techniques and prevention strategies. In:

Abu Dhabi international petroleum exhibition & conference. SPE-192784-MS. URL: https://doi.org/10.2118/192784-MS

[44] Vlasov V.V. (2003) The effectiveness of the use of a standard rod pump in the processes of pumping multicomponent liquid // Oil and gas business. – No. 2 – pp. 1-7

[45] Geological report on the Allagulovskoye field for 2010. – 38 p.

[46] Ivanova L.V. (2011) Asphaltosmoloparaffin deposits in the processes of extraction, transport and storage / L.V. Ivanova, E.A. Burov, V.N. Koshelev // Electronic scientific journal "Oil and gas business". – No. 1. – pp. 268-284. URL: http://www.ogbus.ru/authors/IvanovaLV/IvanovaLV_1.pdf.

[47] Budarova O.P., Boldyrev S.V. (2020) The Wear of a Piston–Sleeve Friction Pair in Axial-Piston Pumps under the Conditions of Water-Contaminated Lubricating Oil. J. Frict. Wear 41, 31-35. URL: https://doi.org/10.3103/S1068366620010043

[48] Sotoodeh K. (2021) Analysis and Failure Prevention of Nozzle Check Valves Used for Protection of Rotating Equipment Due to Wear and Tear in the Oil and Gas Industry. J Fail. Anal. and Preven. 21, 1231-1239. URL: https://doi.org/10.1007/s11668-021-01162-2

[49] Kurbanbayev M.I. Improving the efficiency of oil production wells based on the use of mixtures of multifunctional water-soluble surfactant compositions and polymers. Dissertation of Candidate of Technical Sciences, 2011. – 140 p. URL: https://www.dissercat.com/content/povyshenie-effektivnosti-raboty-neftedobyvayushchikh-skvazhin-na-osnove-ispolzovaniya-smesei

[50] Ibragimov N.G., Hafizov A.R., Shaidakov V.V., etc. Complications in oil production, Ufa: Monograph, 2003. – 302 p. URL: https://www.studmed.ru/ibragimov-n-g-hafizov-a-r-shaydakov-v-v-i-dr-oslozhneniya-v-neftedobyche_59a52f12d7f.html

[51] Timonin V.I., Demko T.T. Extraction, collection and in-field transport of high-paraffin oil in the fields of Southern Mangyshlak, M., All-RRIOMEOGI, 1973.

[52] Galonsky P.P. The fight against paraffin in oil production. Theory and practice/ P.P. Galonsky – M.: Gostoptehizdat, 1955. – 151 p.

[53] Shock D.A. Studies of the mechanism of paraffin deposition and its control / D.A. Shock, J.O. Sudbury, J.J. Crockett // JPT. - 1955. – pp. 23-28.

[54] Glushchenko V.N., Silin M.A. Oilfield chemistry. Ed. in 5 volumes. – Vol.3 Bottom-hole formation zone and technogenic factors of its condition. – M.:

Intercontact. – Science, 2010. – 650 p. URL: https://neft-gaz-novacii.ru/ru/bookshop/chemicals

[55] Glushchenko V.N., Silin M.A., Guerin Yu.G. Oilfield chemistry. - Ed. in 5 volumes - M.: Intercontact Science, 2009. – Vol.5 Prevention and elimination of asphaltene–resin–paraffin deposits. – 475 p. URL: https://neft-gaz-novacii.ru/ru/bookshop/chemicals

[56] Shcherbakov G.Yu. Composition for removal of asphaltene-resin-paraffin deposits in producing wells of oil and gas condensate fields on a hydrocarbon basis / G.Yu. Shcherbakov, A.B. Petukhov, G.M. Khalikova // Oil and gas business.T. 13. No. 2. – Ufa, 2015. – pp. 80-83.

[57] Abdelazim A., Abu El Ela M., El-Banbi A. et al. (2022) Successful approach to mitigate the asphaltenes precipitation problems in ESP oil wells. J Petrol Explor Prod Technol 12, 725-741. URL: https://doi.org/10.1007/s13202-021-01335-7

[58] Al-Taq A.A., Zeid S.M.A., Al-Haji H.H., Saleem J.A. (2013) Removal of organic deposits from oil producing wells in a sandstone reservoir: a lab study and a case history. Society of Petroleum Engineers. URL: https://doi.org/10.2118/164410-MS

[59] Al-Taq A.A., Muhaish S.A., Nakhli M.M., Alrustum A.A. (2015) Organic/inorganic deposition in oil-producing wells from carbonate reservoirs: mechanisms, removal, and mitigation. Society of Petroleum Engineers. URL: https://doi.org/10.2118/177447-MS

[60] Savinkin V.V., Ivanova O.V., Dmitriev F.S., Netesova E.A. Problems of efficient operation of downhole deep rod pumps and promising ways of solving technological problems in the production of hydrocarbons. Materials of the scientific journal "Bulletin of M.Kozybayev NKSU" No. 3(36) - Petropavlovsk: M. Kozybayev NKSU, 2017. – pp. 50-55

[61] Chaogang C., Jixiang G., Na A., et al (2012) Study of asphaltene dispersion and removal for high-asphaltene oil wells. Pet. Sci. 9:551-557

[62] Kuanyshbaeva E.Zh., Baymukhametov M.A., Kazangapov A.E. (2018). Evaluation of the efficiency of production well operation in the conditions of application of new technical means. Bulletin of KazNTU named after K.I. Satpayev 4:260-264

[63] Results of pilot tests of Ozenmunaygas JSC and Munaymash JSC for 2017-2018. – 28 p.

[64] Boyko G.I., Lyubchenko N.P., Aitkalieva G.S., Sarmurzina R.G., Karabalin U.S.,

Tiesov D.S., Bukaeva G.A. Asphalt-resin-paraffin deposits of deposits of Western Kazakhstan and compositions for their removal. Bulletin of the Oil and gas industry of Kazakhstan No. 1 (2) 2020. – pp. 60-69

[65]   Macepuro E.A. Metrological support of the research of the salt inhibitor efficiency. "Scientific and Technical Journal "Metrology" No. 4 (67). Astana: Republican State Enterprise "Kazakhstan Institute of Metrology", 2018. – pp.69-73

[66]   Lichinga, K.N., Luanda, A. & Sahini, M.G. (2022) A novel alkali-surfactant for optimization of filtercake removal in oil–gas well. J Petrol Explor Prod Technol 12, 2121–2134. URL: https://doi.org/10.1007/s13202-021-01438-1

[67]   GOST 2477-2014 (2018) Interstate standard. Oil and petroleum products. Method for determining the water content. Official edition. – Moscow: Standartinform, 12 p. URL: https://files.stroyinf.ru/Data2/1/4293763/4293763540.pdf

[68]   Types and classification of underground work in wells // studopedia.info. URL: https://studopedia.info/5-64511.html

[69]   Sharifullin A.M. The method of repair of the borehole rod deep pump. Patent RF, RU   2282750   C1,   Bul.   No.   24,   27   August   2006.   URL: https://patents.google.com/patent/RU2282750C1

[70]   Ziyatdinov R.Z. The method of repair of the production column of the producing well. Patent RF, RU 2730158 C1, Bul. No. 23, 22 April 2020. URL: https://patents.google.com/patent/RU2730158C1

[71]   Muradkhanov I.V. Oil and gas field equipment. – Stavropol: North Caucasus Federal University, 2016. – pp. 13-18.

[72]   Rocking machine // rengm.ru URL: https://rengm.ru/rengm/stanok-kachalka.html

[73]   Baidildina O.Zh., Daribaeva N.G., Kurmanbaeva B.M. (2015) Features of the structure and properties of paraffinic oils of Kazakhstan, affecting the effectiveness of measures to combat paraffin deposition // Modern high-tech technologies. – No. 4 – pp. 100 – 106.

[74]   Mishchenko I.T., Sakharov V.A., Gron V.G., Bogomolny G.I. (2015) Collection of tasks on oil production technology and technique. – Tomsk: Tomsk Polytechnic University Publishing House. – 68 p.

[75]   Swarovskaya N.A. (2009) Chemistry of oil and gas. – Tomsk: Tomsk Polytechnic University Publishing House. – 111 p.

[76] Mohs Hardness Scale // modificator.ru URL: http://www.modificator.ru/terms/hardness_mohs.html

[77] Zakharov, B.S., Sharikov, G.N. & Kormishin, E.G. (2004) Updating piston pumps for oil production. Chem Petrol Eng 40, 732–738. URL: https://doi.org/10.1007/s10556-005-0041-6

[78] Blinov Yu.I., Klimov V.P., Kuznetsov V.I., Pykhov S.I., Fedorin V.R, Chernyshevich S.L., Shurinov V.A., Khokhlov N.P., Yarygin V.N. Collar plunger of submerged well pump. Patent RF, RU 2096661 C1, 20 November 1997. URL: https://patents.google.com/patent/RU2096661C1

[79] Shurinov V.A., Pykhov S.I., Kozlovsky A.M., Bezzubov A.V. (1999) Oil-well sucker-rod pump. Patent RF, RU 2140571 C1, 27 October 1999. URL: https://patents.google.com/patent/RU2140571C1

[80] Solyanikov B.G., Lorenz F.F., Belyaev S.N., Kolevatov V.B., Seleznev Yu.M. (1998) Method of making pump rods. Patent RF, RU 2119858 C1, 10 October 1998. URL: https://patents.google.com/patent/RU2119858C1

[81] Valiakhmetov O.R., Mulyukov R.R., Gazarov A.G. (2014) Method of pump rods production for deep-well pump. Patent RF, RU 2329129, 09 October 2014. URL: https://patents.google.com/patent/RU2329129C2

[82] Heinrich Richmuller. Oil extraction by deep rod pumps. Publishing house "Scheler-Blackman GMBH", Terlitz, 1988. – p.150

[83] Gedz A.D., Kibernichenko A.B., Efremov V.V., Gilevsky R.V. (2005) Installation for gas-flame spraying and abrasive-jet surface preparation. Patent RF, RU43549U1, 27 January 2005. URL: https://patents.google.com/patent/RU43549U1

[84] Buhl H. (2001) Korrosion und Korrosionsschutz: Titan und Titanlegierungen, Zirconium, Tantal und Niob. Weinheim, Wiley-VCH Verlag GmbH, p.1370-1486

[85] Aksenov A.G., Ermolaev A.S., Muratov K.R. et al. (2022) Laser Powder Surfacing of Gas-Turbine Components and Restoration of the Metal Powder. Russ. Engin. Res. 42, 860-862. URL: https://doi.org/10.3103/S1068798X22080068

[86] Tominaga Y., Kim J., Pyun Y. et al. (2013) A study on the restoration method of friction, wear and fatigue performance of remanufactured crankshaft. J Mech Sci Technol 27, 3047-3051. URL: https://doi.org/10.1007/s12206-013-0824-1

[87] Belkin P.N., Kusmanov S.A. (2021) Plasma Electrolytic Carburising of Metals and Alloys. Surf. Engin. Appl. Electrochem. 57, 19-50. URL:

https://doi.org/10.3103/S1068375521010038

[88] Kartsev S.V. (2020) Mathematical Model of Optimization of Controlled Parameters of the Plasma Surfacing Technological Process of Wear-Resistant Coatings. J. Mach. Manuf. Reliab. 49, 823-828. URL: https://doi.org/10.3103/S1052618820090095

[89] Niu X., Singh S., Garg A. et al. (2019) Review of materials used in laser-aided additive manufacturing processes to produce metallic products. Front. Mech. Eng. 14, 282-298. URL: https://doi.org/10.1007/s11465-019-0526-1

[90] Savinkin V.V., Vizureanu P., Sandu A.V., Ivanischev A.A., Surleva A. (2020) Improvement of the turbine blade surface phase structure recovered by plasma spraying. Coatings 10(1):62. https://doi.org/10.3390/coatings10010062

[91] Savinkin V.V., Kuznetsova V.N., Ratushnaya T.Yu., Kiselev L.A. (2019) Method of integrated assessment of fatigue stresses in the structure of the restored blades of CHP and HPS. Bulletin of the Tomsk Polytechnic University, Geo Assets Engineering 330(8):65-77. https://doi.org/10.18799/24131830/2019/8/2213

[92] Koval A.D., Shmyrko V.I. Complex studies of the properties of heat-resistant alloys ZHS6U, VZHL12U, EP957VD, ZMI-3, ZMI-ZU. Bulletin of Engine Building, No. 1/2006. Structural materials. – p.143-146. URL: file:///C:/Users/User/Downloads/kompleksnye-issledovaniya-svoystv-zharoprochnyh-splavov-zhs6k-vzhl-12u-ep-957vd-zmi-3-zmi-zu%20(1).pdf

[93] Som A.I. Iron-based alloy for plasma-powder surfacing of screw extruders and injection molding machines // Automatic welding, №7 (754), 2016. – pp. 22-27. URL: https://patonpublishinghouse.com/as/pdf/2016/pdfarticles/07/5.pdf

[94] Kolisnichenko S.N., Koptyaev D.A., Kolisnichenko S.V. Investigation of energy-efficient technologies and technological complexes for restoring blades of complex geometry of power equipment, using plasma energy. The development of the forms of the method of management, economics and law in the minds of globalizatsiy: Materials of the 4th National Sciences.-practice. conf. – Dnipro, 5-7.04.2016: tezi dopovidey. – Vol.1. – Dnipro: Vidavnitstvo "Svidler A.L.", 2016. – pp. 307-309

[95] Gazarov A.G. (2004) Development of methods to reduce the wear of rod pumping equipment in deviating wells. Abstract, Ufa, p. 24

[96] Nagirny V.M., Prikhodko L.A., Vetlyanskaya T.V. (1993) Method for strengthening articles from nickel or its alloys. Patent RF, RU 1820916 C, 07 June

1993. https://patents.google.com/patent/RU1820916C

[97] Frenkel Ya.I. (2021) Introduction to the theory of metals. URSS. Series: Physical and Mathematical heritage: Physics (Solid State Physics). Series: Classics of Engineering: Metal Science, p. 432

[98] Gorunov A.I. (2017) Restoration of Aircraft Gas Turbine Engine Titanium Compressor Blades by Laser Surfacing. Metallurgist 61, 498–504. URL: https://doi.org/10.1007/s11015-017-0523-8

[99] Nasteka D.V., Grachev O.E., Silevich V.M. et al. (2019) Restoration of Steam Turbine Last Stage Rotor Blades by Laser Surfacing. Power Technol Eng 53, 208-212. URL: https://doi.org/10.1007/s10749-019-01061-5

[100] Ahn D.G. (2021) Directed Energy Deposition (DED) Process: State of the Art. Int. J. of Precis. Eng. and Manuf.-Green Tech. 8, 703-742. URL: https://doi.org/10.1007/s40684-020-00302-7

[101] Patterson T., Hochanadel J., Sutton S. et al. (2021) A review of high energy density beam processes for welding and additive manufacturing applications. Weld World 65, 1235-1306. URL: https://doi.org/10.1007/s40194-021-01116-0

[102] Meng Y., Xu J., Ma L. et al. (2022) A review of advances in tribology in 2020–2021. Friction 10, 1443-1595. URL: https://doi.org/10.1007/s40544-022-0685-7

[103] Kachurin A.E., Beketov S.B. Features of well operation, equipped with rod pumps in fields with weakly cemented reservoirs // Mining Information and Analytical Bulletin (Scientific and Technical Journal), 2010, p.107-115

[104] Arbuzov V.N. Operation of oil and gas wells: a textbook. Part 2 / V.N. Arbuzov; Tomsk Polytechnic University. – Tomsk: Publishing House of Tomsk Polytechnic University, 2012. – p. 272

[105] Ivanova L.V. Influence of chemical composition and water content of oil on the amount of asphalt-resin-paraffin deposits. / Ivanova L.V., Vasechkin A.A., Koshelev V.N. // Petrochemistry, 2011. Vol. 51. No. 6. – pp. 403-409.

[106] Ivanova L.V. An integrated approach to the study of the group and structural-group composition of the oil fields of Udmurtia. / Ivanova L.V., Miller V.K., Primerova O.V., Koshelev V.N., Burov E.A. // Butlerov messages. 2014. Vol. 39. No. 8. – pp. 50-56.

[107] Galimullin M.L. (2000) Improving the reliability of rod depth pumps. Scientific problems of the Volga-Ural Oil and Gas region. Technical and natural aspects 2:25-28

[108] Zirconium dioxide, aluminum oxide and their compounds // virial.ru. URL: http://www.virial.ru/materials/95/

[109] Edison I.A. Development of technology for obtaining nanopowders of aluminum and zirconium oxides and materials based on them by spray drying of solutions and suspensions: dis. Candidate of Technical Sciences: 05.17.11 / National Research Tomsk Polytechnic University. – Tomsk, 2020. – 163 p.

[110] ATmega328P [DATASHEET] 8-bit AVR Microcontroller with 32K Bytes In-System Programmable Flash. – 2015: Atmel Corporation. – 294 p.

[111] Arduino Nano (V2.3) User Manual. – 2008: Creative Commons Attribution Share-Alike 2.5. – 5 p.

[112] Veiko V.P., Shakhno E.A. Collection of problems on laser technologies. – St. Petersburg: ITMO State University, 2007. – 67 p.

[113] Power Density Calculator // ophiropt.com/ URL: https://www.ophiropt.com/laser--measurement/ru/power-density-calculator

[114] Zirconium Oxide balls (ZrO$_2$) // rgpballs.com/ URL: https://www.rgpballs.com/en/zirconium-oxide-zro2-balls/

[115] QBH Fiber Optic Cable: 1030 nm to 1090 nm Datasheet. – 2022: Coherent, Inc. – 7 p.

[116] Alkhimov A.P. Scientific foundations of the technology of cold gas-dynamic spraying (CGDS) and properties of sprayed materials: monograph / A.P. Alkhimov, V.F. Kosarev, A.V. Plokhov. – Novosibirsk: NSTU Publishing House, 2006. – 280 p. – ("Monographs of NSTU").

[117] Molodyk N.V., Zenkin A.S. Restoration of machine parts. Handbook. – M.: Mechanical Engineering, 1989. – 480 p.

[118] Panteleenko F.I., Lyalyakin V.P., Ivanov V.P., Konstantinov V.M.; Restoration of machine parts. Handbook. – M.: Mechanical Engineering, 2003. – 672 p.

[119] Edited by Borisov Yu.S., Noskin R.A. Handbook of mechanics of a machine-building plant. Guide. in 2 t. supplemented – M.: Mechanical Engineering, 1971. – 565 p.

www.ingramcontent.com/pod-product-compliance
Lightning Source LLC
Chambersburg PA
CBHW071710210326
41597CB00017B/2414